Christoph Haefeli

Finite volume effects in Chiral Perturbation Theory

Christoph Haefeli

Finite volume effects in Chiral Perturbation Theory

Meson masses and decay constants

Südwestdeutscher Verlag für Hochschulschriften

Impressum/Imprint (nur für Deutschland/ only for Germany)
Bibliografische Information der Deutschen Nationalbibliothek: Die Deutsche Nationalbibliothek verzeichnet diese Publikation in der Deutschen Nationalbibliografie; detaillierte bibliografische Daten sind im Internet über http://dnb.d-nb.de abrufbar.
Alle in diesem Buch genannten Marken und Produktnamen unterliegen warenzeichen-, marken- oder patentrechtlichem Schutz bzw. sind Warenzeichen oder eingetragene Warenzeichen der jeweiligen Inhaber. Die Wiedergabe von Marken, Produktnamen, Gebrauchsnamen, Handelsnamen, Warenbezeichnungen u.s.w. in diesem Werk berechtigt auch ohne besondere Kennzeichnung nicht zu der Annahme, dass solche Namen im Sinne der Warenzeichen- und Markenschutzgesetzgebung als frei zu betrachten wären und daher von jedermann benutzt werden dürften.

Verlag: Südwestdeutscher Verlag für Hochschulschriften Aktiengesellschaft & Co. KG
Dudweiler Landstr. 99, 66123 Saarbrücken, Deutschland
Telefon +49 681 37 20 271-1, Telefax +49 681 37 20 271-0, Email: info@svh-verlag.de
Zugl.: Bern, Universität, Diss., 2005

Herstellung in Deutschland:
Schaltungsdienst Lange o.H.G., Berlin
Books on Demand GmbH, Norderstedt
Reha GmbH, Saarbrücken
Amazon Distribution GmbH, Leipzig
ISBN: 978-3-8381-0726-4

Imprint (only for USA, GB)
Bibliographic information published by the Deutsche Nationalbibliothek: The Deutsche Nationalbibliothek lists this publication in the Deutsche Nationalbibliografie; detailed bibliographic data are available in the Internet at http://dnb.d-nb.de.
Any brand names and product names mentioned in this book are subject to trademark, brand or patent protection and are trademarks or registered trademarks of their respective holders. The use of brand names, product names, common names, trade names, product descriptions etc. even without a particular marking in this works is in no way to be construed to mean that such names may be regarded as unrestricted in respect of trademark and brand protection legislation and could thus be used by anyone.

Publisher:
Südwestdeutscher Verlag für Hochschulschriften Aktiengesellschaft & Co. KG
Dudweiler Landstr. 99, 66123 Saarbrücken, Germany
Phone +49 681 37 20 271-1, Fax +49 681 37 20 271-0, Email: info@svh-verlag.de

Copyright © 2009 by the author and Südwestdeutscher Verlag für Hochschulschriften Aktiengesellschaft & Co. KG and licensors
All rights reserved. Saarbrücken 2009

Printed in the U.S.A.
Printed in the U.K. by (see last page)
ISBN: 978-3-8381-0726-4

Contents

Overview   1

I   An asymptotic formula for the pion decay constant in a large volume   7

1 Introduction   9

2 The asymptotic formula   10

3 Outline of the derivation   11

4 The coupling constant G_π 14

5 The asymptotic formula in Chiral Perturbation Theory   15

6 Conclusions   18

    Acknowledgments   19

    References   20

II   Finite volume effects for meson masses and decay constants   21

1 Introduction   23

2 Finite volume effects   25
    2.1 ChPT in finite volume   25
    2.2 Lüscher formula   27

3 The Lüscher formula resummed   29

4 Meson masses and decay constants in finite volume   32
    4.1 Pion   33
       4.1.1 Pion mass   33

4.1.2 Pion decay constant	34
4.2 Kaon	35
4.2.1 Kaon mass	36
4.2.2 Kaon decay constant	36
4.3 Eta	37

5 Summary of the analytical results   38
    5.1 Full formulae   38
    5.2 Simplified formulae   41

6 Numerical analysis   42
    6.1 M_π dependence of F_π 43
    6.2 M_π and m_s dependence of M_K, F_K and M_η 43
    6.3 Results   44

7 Two types of applications   52
    7.1 Finite volume effects and Marciano's determination of V_{us} 52
    7.2 Low energy constants from finite volume effects   54

8 Conclusions   58

    Acknowledgments   59

A The integrals $S_{M_P}^{(n)}$ and $S_{F_P}^{(n)}$ 59

B Cut-off effects   64

C Effects due to kaon and eta loops   66

    References   68

III Finite volume effects for the pion mass at two-loops   71

1 Introduction   73

2 Preliminaries   75
    2.1 ChPT in finite and in infinite volume   75
    2.2 Basic definitions   77

3 Outline of calculation and statement of results   77
    3.1 One-loop result   77
    3.2 Minimal set of periodified propagators   79

3.3 Two-loop result	80
3.4 Large L limit	81
3.5 Self-energy to 0'th order: $\Sigma^{(0)}$	81
3.6 Self-energy to 1'st order: $\Sigma^{(1)}$	82
3.7 Self-energy to 2'nd order: $\Sigma^{(2)}$	84
4 Summary of analytical results	**87**
5 Numerics	**88**
5.1 Setup	88
5.2 Results	88
6 Summary	**90**
A Finite volume integrals	**93**
A.1 Tadpole	93
A.2 Sunset	94
References	97

IV The pion mass in finite volume to two loops   99

1 Introduction   **101**

2 ChPT in finite volume   **102**

3 Pion in finite volume   **103**
    3.1 Self-energy to 0'th order: $\Sigma^{(0)}$ 104
    3.2 Self-energy to 1'st order: $\Sigma^{(1)}$ 104
    3.3 Self-energy to 2'nd order: $\Sigma^{(2)}$ 105

4 Numerics   **105**

5 Conclusions   **106**

    Acknowledgments   107

    References   108

Overview

It appears to be amazing that the whole range of strong interaction phenomena can be described by quantum chromodynamics (QCD), a theory with only few parameters. Of course we can not be sure whether this is really the case, but the accumulated evidence strongly suggests this to be so. On the experimental side, the masses of the light hadrons and many other properties are very precisely known. On the theoretical side, it is difficult to compute these quantities from first principles, mainly because the coupling constant is not small at low energies and a perturbative expansion in this parameter can not be carried through. Still, it is of great importance to verify that the same lagrangian which describes the interactions between quarks and gluons at high energies also explains the spectrum of light hadrons correctly.

The framework of lattice QCD is appropriate to address this problem on a non-perturbative level. In this framework, one introduces a finite space-time lattice and discretizes the fields. Quantitative results can then be obtained through sophisticated numerical simulations. One should however not conclude that the theory can be solved through numerical simulations alone. Due to the limited capacity of today's computer systems, the lattices one can simulate are rather small and coarse and the simulated quark masses are slightly heavier than in nature. To give a rough estimate, present lattice calculations (with dynamical fermions) performing spectroscopy are limited to lattice sizes of about $L = 2$fm, at a lattice spacing of about $a = 0.1$fm. Hadrons contained in such small volumes occupy a significant fraction of the available space and one therefore expects that finite-size (as well as cut-off) effects alter the mass of the hadron to a certain extent. Before one can make a comparison with experimentally measured quantities, one has to make three extrapolations, $a \to 0$, $1/L \to 0$ and $m \to 0$, with m the quark mass. They are however by no means straightforward as they involve non-trivial functions of these parameters. Analytical methods may help in this respect. A recent overview of the present status of the analytical calculations related to the extrapolation to the continuum and the chiral limit has been given by Bär [1]. Here, we will discuss the extrapolation to the infinite volume limit.

The physical origin of the size dependence of the masses and decay constants that have been evaluated in this work is due to polarisation effects. Particles polarise the vacuum around them, i.e. they are accompanied by a cloud of virtual particles whose extension is roughly equal to the Compton wave length of the lightest particle, the pion. As soon as the cloud is squeezed by the box, we expect that properties of the particles start to show a size dependence. And since the effect mainly comes from the cloud and not from the particle itself, it should be of the same order for eg. the nucleon mass and the pion mass. In fact, the quantities start to show a size dependence even for slightly larger volumes. They are proportional to $\exp(-M_\pi L)$, (with M_π the mass of the pion in infinite volume) and become very small as soon as the size of the box is about three times larger than the diameter of the polarisation cloud.

This physical picture can be translated in a quantitative formulation. In the presence of a large, but finite volume chiral symmetry dictates the dynamics of the pions at low energies, and one can study them systematically in the framework of Chiral Perturbation Theory (ChPT)[2]. As lattice calculations approach smaller quark masses and aim at higher precision, analytical studies of finite-size effects will become of increasing importance. In the past three years while this thesis was being worked out, a number of finite-size studies in different quantities appeared in the literature [3, 4]. One of these influenced the thesis fruitfully [3]. The authors evaluated finite size-effects of the pion mass beyond leading order and found substantial subleading effects, of the order of 50% with respect to the leading contributions, even for small values of the quark masses. The comparison of next-to-leading with next-to-next-to-leading however shows that the chiral expansion does have a good converging behaviour. This accurate study of the convergence of the chiral series has been made possible by the use of Lüscher's asymptotic formula for the pion mass [5],

$$M_{\pi L} - M_\pi = -\frac{3}{16\pi^2 M_\pi L} \int_{-\infty}^{\infty} dy\, \mathcal{F}_\pi(iy)\, e^{-\sqrt{M_\pi^2 + y^2}L} + O(e^{-\bar{M}L})\,. \qquad (1)$$

The formula relates its leading finite-size corrections to an integral over the forward $\pi\pi$ scattering amplitude $\mathcal{F}_\pi(\nu)$ in infinite volume and $\bar{M} \geq \sqrt{2}M_\pi$. Since the latter is known to next-to-next-to-leading order in the chiral expansion [6], the Lüscher formula could be evaluated to the same order in the chiral expansion. In view of the results for the pion mass, the question arose, whether one can derive similar asymptotic formulae also for other quantities.

In the first part of this work, "An asymptotic formula for the pion decay constant in a large volume", published in PLB 590 (2004) 258, we show that this is indeed the case and derive an asymptotic formula for decay constants. In case of the pion decay constant, the formula is expressed as an integral over a subtracted $\langle 3\pi|A_\mu|0\rangle$ amplitude. The amplitude has a pole due to the direct coupling of the axial current to a pion, which then rescatters into three pions. This pole appears exactly in the kinematical region where it is needed and must be subtracted. We discussed the prescription of the subtraction as well as its physical interpretation and analysed the formula numerically at leading and next-to-leading order in the chiral expansion. The results showed again large next-to-leading order corrections.

Most of the recent studies in [4] relied on a one-loop calculation in ChPT and not on asymptotic formulae. In the case of the pion mass the one-loop result was presented in [7] and reads

$$M_{\pi L} = M_\pi \left[1 + \frac{1}{4}\xi \tilde{g}_1(\lambda) + O(\xi^2)\right]\,, \qquad (2)$$

with $\lambda = M_\pi L$, $\xi = (M_\pi/4\pi F_\pi)^2$ and

$$\tilde{g}_1(\lambda) = \sum_{n=1}^{\infty} \frac{4m(n)}{\sqrt{n\lambda}} K_1(\sqrt{n}L) , \tag{3}$$

($m(n)$ is the multiplicity of a vector r with $r^2 = n$). It is instructive to compare the result of the one-loop calculation (2) with the asymptotic formula (1). If one inserts the tree level chiral representation for the $\pi\pi$ scattering amplitude into eq. (1),

$$\mathcal{F}_\pi(\nu) = -\frac{M_\pi^2}{F_\pi^2} + \mathcal{O}(M_\pi^4) , \tag{4}$$

one obtains the same expression for the leading exponential term as in eq.(2), but the asymptotic formula misses the terms apart from $n = 2$ in the series of eq.(2). It takes into account only the leading exponential contributions of the order $\exp(-M_\pi L)$ and systematically drops contributions of the order $\exp(-\sqrt{2}M_\pi L)$. The comparison to the numerical values obtained with the full one-loop ChPT formula eq.(2) shows that for moderate values of $M_\pi L$ exponentially subleading terms may be important. As has been noted by Colangelo in [8], there is actually a straightforward extension of the asymptotic formula through a resummation which then reproduces the whole series of the one-loop result in eq.(2). The conjecture is that this resummation takes into account the most important part of higher order contributions. This is not obvious, because the resummed asymptotic formula misses various terms starting at the two-loop order. Only a complete two-loop calculation can show how big these contributions are numerically. In the third part of this work, "Finite volume effects of the pion mass at two-loops", this conjecture is investigated with a full two-loop calculation of the pion mass within the framework of ChPT.[1] The results of this calculation confirm our expectation that contributions which were omitted in the resummed asymptotic formula are very small for $M_\pi L \gtrsim 2$ at the two-loop level.

We make some further comments about the two-loop calculation compared to the asymptotic formula. It is well known that finite-size effects do not alter the uv-behaviour of the theory (see eg. [2]). They are purely infrared effects and no further effective couplings have to be introduced to regularise the theory. Still, when evaluating finite-size effects perturbatively in terms of Feynman diagrams, infrared and ultraviolet contributions intermix at intermediate stages and one has to find an economical way in order to separate the uv-divergences. The asymptotic formula does not involve such difficulties, the finite volume shift is expressed in terms of renormalised quantities and to draw level with the two-loop calculation, only a one-loop calculation of a scattering amplitude in infinite volume has to be performed.

---

[1] Preliminary results of this calculation were reported at the XXIII International Symposium of Lattice Field Theory in Dublin, Ireland. In part IV of this work we enclose the proceedings of this talk.

The asymptotic formula for the pion mass is preferable with respect to a full two-loop calculation for another reason: it represents the finite-size effects in a very compact manner. The exponential suppression factor $\exp[-\sqrt{M_\pi^2 + y^2}L]$ in eq. (1) makes apparent that the most important numerical contributions are obtained in the vicinity of $y = 0$. This can be used to simplify the asymptotic formula considerably. In the immediate vicinity of $y = 0$, a polynomial approximation for the chiral amplitudes reproduces them rather accurately. The advantage of such a polynomial representation is that all integrals can be performed analytically, leading to a handy and convenient representation of the finite-size effects. We doubt that the knowledge of only the full two-loop formulae would have led us to such a simplified representation. Still, the calculation showed that there are no principal limitations to carry through a two-loop calculation in finite volume for yet other quantities, where subleading finite-size effects might turn out to be important too, but are not accessible with an asymptotic formula.

The discussion above has shown the usefulness of the asymptotic formulae. In the second part of this work, " Finite volume effects for meson masses and decay constants", published in NPB 721 (2005) 136, we have applied the resummed asymptotic formula to the whole pseudoscalar octet and provide analytical finite-size formulae for M_π, F_π, M_K, F_K and M_η. For the smallest acceptable values of $M_\pi L \gtrsim 2$ and $L \gtrsim 2$fm, we find that the finite-size effects are of the order of a few percent for M_π, F_π and F_K, whereas the masses M_K and M_η are practically insensitive to the box size.

The analytical formulae allowed us to discuss two types of applications which we find worthwhile to be noted here. The first one concerns the determination of low energy constants from finite-size effects: the asymptotic formulae provide a connection between finite-size effects on two-point functions and (infinite volume) four-point functions, and thus give us access to low energy constants that appear as local contributions to four-point functions. Due to the fact that our finite-size effects are exponentially suppressed and numerically quite small, we find that one would have to control the pion mass and the pion decay constant for eg. $M_\pi = 300$ MeV and $L = 2$ fm to less than 1 permille in order to get a reasonable account on the (combination of) low energy constants, which is a real challenge – but might be worthwhile to explore compared to the difficult calculation with the alternative method of Lüscher [9].

The second application that we would like to mention is an example of how the analytic finite-size formulae may help to control a systematic error in lattice calculations. In [10], Marciano suggested to determine the CKM matrix element V_{us} via the phenomenologically known branching ratio for $K_{\ell 2}$ and $\pi_{\ell 2}$, the CKM matrix element V_{ud} and a lattice calculation of the ratio F_K/F_π. The necessary accuracy to make an impact on the determination of V_{us} is at the level of 1% or better, and indeed, both the determination of V_{ud} as well as the ratio of branching ratios are known to well below 1%. This means that any improvement in the lattice calculation of F_K/F_π will be immediately reflected in the value of V_{us}. In particular, being able to control systematic effects to well below 1% is of crucial importance. With our results for F_K and F_π, it is straightforward to calculate the finite-size shift of the ratio F_K/F_π and thus to compare the magnitude of this effect to the typical size of the statistical error.

In summary, this thesis presents a detailed analysis of finite-size effects for masses and decay constants of the octet of pseudoscalar mesons. Thereby, we have relied to a large extent on asymptotic formulae à la Lüscher combined with the chiral representation for the corresponding scattering amplitudes. The cost of a lattice calculation grows very rapidly with the box size – in case of M_π, F_π, M_K, F_K and M_η our results show that it is unnecessary to make the infinite-size extrapolation numerically, and one can correct for finite-size effects analytically.

References

[1] O. Bar, Nucl. Phys. Proc. Suppl. 140, 106 (2005) [hep-lat/0409123].

[2] J. Gasser and H. Leutwyler, Nucl. Phys. B 307, 763 (1988).

[3] G. Colangelo, S. Durr and R. Sommer, Nucl. Phys. Proc. Suppl. 119, 254 (2003) [hep-lat/0209110]. G. Colangelo and S. Durr, Eur. Phys. J. C 33, 543 (2004) [hep-lat/0311023].

[4] D. Becirevic and G. Villadoro, Phys. Rev. D 69, 054010 (2004) [hep-lat/0311028]. A. Ali Khan et al. [QCDSF-UKQCD Collaboration], Nucl. Phys. B 689, 175 (2004) [hep-lat/0312030]. D. Arndt and C. J. D. Lin, Phys. Rev. D 70, 014503 (2004) [hep-lat/0403012]. S. R. Beane, Phys. Rev. D 70, 034507 (2004) [hep-lat/0403015]. S. R. Beane and M. J. Savage, Phys. Rev. D 70, 074029 (2004) [hep-ph/0404131]. Y. Koma and M. Koma, Nucl. Phys. B 713, 575 (2005) [hep-lat/0406034]. Y. Koma and M. Koma, [hep-lat/0504009]. A. Fuhrer, The nucleon in finite volume, Master Thesis, Universität Bern (2004). Can be obtained from
http://www.itp.unibe.ch/index.html?lang=0&id=2&subsubid=0 .

[5] M. Lüscher, Commun. Math. Phys. 104, 177 (1986).

[6] J. Bijnens et al., Phys. Lett. B 374 210 (1996) [hep-ph/9511397], Nucl. Phys. B 508 263 (1997) [Erratum-ibid. B 517 639 (1998)] [hep-ph/9707291].

[7] J. Gasser and H. Leutwyler, Phys. Lett. B 184, 83 (1987).

[8] G. Colangelo, Nucl. Phys. Proc. Suppl. 140, 120 (2005) [hep-lat/0409111].

[9] M. Luscher, Commun. Math. Phys. 105, 153 (1986).

[10] W.J. Marciano, Phys. Rev. Lett. 93, 231803 (2004) [hep-ph/0402299].

I

An asymptotic formula for the pion decay constant in a large volume

published in

Phys. Lett. B 590 (2004) 258

An asymptotic formula for the pion decay constant in a large volume

Gilberto Colangelo and Christoph Haefeli

Institut für Theoretische Physik, Universität Bern
Sidlerstr. 5, 3012 Bern, Switzerland

Abstract

We derive an asymptotic formula à la Lüscher for the finite volume correction to the pion decay constant: this is expressed as an integral over the $\langle 3\pi|A_\mu|0\rangle$ amplitude after proper subtraction of the pion pole contribution. We analyze the formula numerically at leading and next-to-leading order in the chiral expansion.

1 Introduction

The analytical study of finite volume effects is becoming of increasing importance as lattice calculations with dynamical fermions approach smaller quark masses and aim at higher precision. Since these effects are dominated by the lightest particles in the spectrum, the pions, and by their long distance dynamics, one can study them in the framework of chiral perturbation theory (CHPT) [1]. A number of analyses of these effects in different quantities have recently appeared in the literature [2, 3]. One of these concerned the case of the pion mass [2] and has shown that a leading order calculation may receive very large corrections from the next-to-leading contribution even for small values of the quark masses, whereas even higher order corrections behave according to expectations and show a convergent behaviour. This accurate study of the convergence of the chiral series has been made possible by the use of Lüscher's asymptotic formula for the pion mass [4]. The formula relates its leading finite-volume corrections to an integral over the $\pi\pi$ scattering amplitude in infinite volume. Since the latter is known to next-to-next-to-leading order in the chiral expansion [5], it is straightforward to evaluate Lüscher's formula to the same order in the chiral expansion.

In view of the results for the pion mass, the question arises if one can derive similar asymptotic formulae also for other quantities: as we will show in what follows, this is the case. In the present article we concentrate on F_π, derive an asymptotic formula which relates it to the infinite-volume $\langle 3\pi|A_\mu|0\rangle$ amplitude and analyze it numerically using the next-to-leading order calculation of this amplitude [6]. The results show again large next-to-leading order corrections – in this case we cannot explore the chiral expansion further because the two-loop calculation of the $\langle 3\pi|A_\mu|0\rangle$ amplitude is not yet available.

2 The asymptotic formula

Denote by $F_{\pi,L}$ the pion decay constant in a box of size L. The asymptotic formula for $\Delta F_\pi = F_{\pi,L} - F_\pi$ can then be written as:

$$\Delta F_\pi = \frac{3}{8\pi^2 M_\pi L} \int_{-\infty}^{\infty} dy\, e^{-\sqrt{M_\pi^2+y^2}L} N_F(iy) + O(e^{-\bar{M}L}) \,, \qquad (1)$$

where $\bar{M} \geq \sqrt{3/2}\, M_\pi$ and the amplitude $N_F(\nu)$ is defined as follows. Consider the amplitude for creation of three pions out of the vacuum with the axial current:

$$\begin{aligned}
\langle \pi^1(p_1)\pi^1(p_2)\pi^3(p_3)|A_\mu^3(0)|0\rangle &= (p_1+p_2)_\mu G(s_1,s_2,s_3) \qquad (2)\\
&+ (p_1-p_2)_\mu H(s_1,s_2,s_3) + p_{3\,\mu} F(s_1,s_2,s_3) \,,
\end{aligned}$$

where the superscripts on the pion states and axial current are isospin indices and G, H and F are three scalar amplitudes of the variables s_1, s_2 and s_3, with $s_1 = (p_2+p_3)^2$ and cyclic permutations [6]. From the amplitude (2) one can construct the combination which has two of the outcoming pions in an $I=0$ state (the explicit relation is given below)

$$\begin{aligned}
\langle (2\pi)_{I=0}\pi^3(p_3)|A_\mu^3(0)|0\rangle &= (p_1+p_2)_\mu G_0(s_1,s_2,s_3) \qquad (3)\\
&+ (p_1-p_2)_\mu H_0(s_1,s_2,s_3) + p_{3\,\mu} F_0(s_1,s_2,s_3) \,.
\end{aligned}$$

This amplitude contains a pole in the unphysical region, for $(p_1+p_2+p_3)^2 = Q^2 = M_\pi^2$, which needs to be removed before proceeding further. We define

$$\begin{aligned}
\langle (2\pi)_{I=0}\pi^3(p_3)|A_\mu^3(0)|0\rangle_S &= \langle (2\pi)_{I=0}\pi^3(p_3)|A_\mu^3(0)|0\rangle \qquad (4)\\
&- Q_\mu \frac{iF_\pi T^{I=0}(s_3, s_1-s_2)}{M_\pi^2 - Q^2} \,,
\end{aligned}$$

where $T^{I=0}(s, t-u)$ is the $\pi\pi$ scattering amplitude with isospin zero in the s channel. We need the subtracted amplitude in the forward kinematic configuration, i.e. for $p_1 = -p_2$, $s_3 = 0$, where it becomes a function of one variable only, $\nu = (s_2 - s_1)/(4M_\pi)$:

$$p_3^\mu \langle (2\pi)_{I=0} \pi^3(p_3) | A_\mu^3(0) | 0 \rangle_S |_{p_1=-p_2} = 2M_\pi \nu h_0(\nu) + M_\pi^2 \bar{f}_0(\nu) \ , \tag{5}$$

where

$$h_0(\nu) = H_0(2M_\pi(M_\pi - \nu), 2M_\pi(M_\pi + \nu), 0)$$

and analogously for $\bar{f}_0$ and where the bar on the F_0 form factor denotes that it is defined after subtraction of the pion pole (the form factor H_0 remains unaffected by the subtraction). The amplitude N_F which enters the asymptotic formula for the finite volume corrections to F_π is defined as

$$N_F(\nu) = -i \left(2\nu h_0(\nu) + M_\pi \bar{f}_0(\nu) \right) \ . \tag{6}$$

The amplitudes H_0 and F_0 can be expressed in terms of F, G and H appearing in (2):

$$\begin{aligned} F_0(s_1, s_2, s_3) &= 3F_{123} + G_{231} + G_{312} - H_{231} + H_{312} \ , \\ H_0(s_1, s_2, s_3) &= 3H_{123} + \frac{1}{2}[F_{231} - F_{312} - G_{231} + G_{312} - H_{231} - H_{312}] \ , \end{aligned} \tag{7}$$

where $X_{ijk} = X(s_i, s_j, s_k)$ with $X = F, G, H$.

3 Outline of the derivation

The derivation of this formula is in large parts analogous to the derivation of the formula for the pion mass, due to Lüscher [4]. In the following we simply outline the necessary steps to prove the formula and refer the reader to the paper of Lüscher for details. The starting point of the analysis is that one can rely on an effective Lagrangian description of the relevant physics and analyze these finite volume effects in CHPT. As observed by Lüscher, the precise form of the effective Lagrangian is never needed in the proof – on the other hand, it is very useful to have it available if one wants to understand in concrete terms these effects. As was shown by Gasser and Leutwyler one can rigorously derive the consequence of

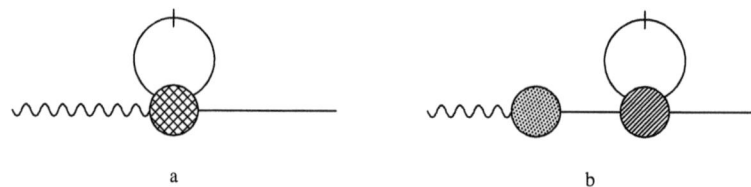

Fig. 1. Graphical representation of the asymptotic formula. The wiggly (straight) line represents the axial current (pion). The dash on the propagator means that it is taken in finite volume (only the contribution with $|\vec{n}| = 1$ in the sum (8)). Diagram a (b) illustrates the correction to F_π (the shift of the pole position) due to finite volume.

chiral symmetry also if the system is closed inside a large finite volume with the help of the effective Lagrangian technique [1]. In particular the form of the local effective Lagrangian remains unchanged, and the only difference with respect to infinite volume calculations comes from the propagator for the pion field which becomes periodic in all spatial directions

$$G(x^0, \vec{x}) = \sum_{\vec{n} \in \mathbb{Z}^3} G_0(x^0, \vec{x} + \vec{n}L) \qquad (8)$$

where $G_0(x)$ is the propagator in infinite volume.

The first step in Lüscher's proof of the asymptotic formula for the pion mass consists in showing that, for a generic loop diagram contributing to the self energy of the pion, the dominating finite volume effect is obtained if one takes all propagators in infinite volume ($G(x) \to G_0(x)$) except one, for which only the terms with $|\vec{n}| = 1$ in the sum in (8) should be kept[1]. The sum of all possible contributions of this form from all possible loop diagrams gives the leading finite volume corrections to the pion mass. The same conclusion is valid also for the Feynman diagrams which renormalize the coupling between the axial current and the pion – the fact that in this case one of the external legs is the axial current instead of a pion does not touch the argument at all.

The second step in the proof consists in showing, by modifying the integration contour in the complex plane, that this leading contribution can be written in a very compact form, as an integral over an amplitude (the $\pi\pi$ scattering amplitude in the case of the pion mass) defined in Minkowski space, analytically continued to complex values of its arguments. Again, the same argument applies also to

---

[1] More precisely: this concerns only propagators which are contained in at least one loop, cf. [4]

the case of the pion decay constant: in this case, in all possible loop graphs that renormalize the pion coupling to the axial current we have to single out one internal pion propagator, break it up and put the resulting two external legs on shell. The relevant amplitude in this case is the $\langle 3\pi | A_\mu | 0 \rangle$ amplitude, as illustrated in fig. 1a – the weight function which appears in the integral is however exactly the same as in the pion mass case.

The kinematic configuration in which the amplitude must be evaluated is also the same and corresponds, for the $\pi\pi$ amplitude, to forward scattering. The $\langle 3\pi | A_\mu | 0 \rangle$ amplitude is however singular for this kinematics because of a pole due to one-pion exchange among the axial current and the three outgoing pions. This singularity does not belong to the finite volume corrections to F_π and should be subtracted. The reason for the presence of this pole can be explained as follows: the $\langle \pi | A_\mu | 0 \rangle$ amplitude is defined as the residue at the pion pole of a two-point function of the axial current and any interpolating field for the pion:

$$\langle \pi^a(q) | A_\mu^b | 0 \rangle = \lim_{q^2 \to M_\pi^2} (M_\pi^2 - q^2) i q_\mu \delta^{ab} P(q) \qquad (9)$$

$$P(q) = N_\phi q^\mu \int dx e^{iqx} \langle 0 | T \phi_\pi^1(x) A_\mu^1(0) | 0 \rangle \ ,$$

with N_ϕ the proper normalization factor which depends on the field ϕ_π. In finite volume both the residue as well as the position of the pole are shifted. Ignoring the latter shift corresponds to multiplying $P_L(q)$ by $(M_\pi^2 - q^2)$ and not by the correct $(M_{\pi,L}^2 - q^2)$ and then taking the limit $q^2 \to M_\pi^2$. The result, expanded to the leading term for asymptotically large volumes, contains a pole for $q^2 = M_\pi^2$

$$(M_\pi^2 - q^2) P_L(q) \sim (M_\pi^2 - q^2) \frac{F_{\pi,L}}{M_{\pi,L}^2 - q^2} = F_{\pi,L} - \frac{F_\pi \Delta M_\pi^2}{M_\pi^2 - q^2} + \ldots \ , \qquad (10)$$

where $\Delta M_\pi^2 = M_{\pi,L}^2 - M_\pi^2$ is also evaluated to leading order. Since the shift in the pion mass is known and given by Lüscher's formula, we can subtract the pole (which is illustrated in fig. 1b) and get the correct finite-volume value of the pion decay constant. The result leads to the subtraction prescription given in the previous section.

4 The coupling constant G_π

The formula presented here for F_π can be extended with obvious modifications also to other quantities, e.g. like G_π, the coupling constant of the pion to the pseudoscalar quark bilinear $P^i = \bar{q} i \gamma_5 \tau^i q$

$$\langle 0|P^i(0)|\pi^k\rangle = \delta^{ik} G_\pi \ . \tag{11}$$

In this case the amplitude that should replace $N_F(\nu)$ in the analogue of eq. (1) is $M_\pi N_G(\nu)$ which is defined through the subtracted $P \to 3\pi$ amplitude in the limit $p_1 = -p_2$:

$$N_G(\nu) = \lim_{p_1 \to -p_2} \left[\langle (2\pi)_{I=0} \pi^3(p_3) | P^3(0) | 0 \rangle - \frac{G_\pi T^{I=0}(s_3, s_1 - s_2)}{M_\pi^2 - Q^2} \right] \ . \tag{12}$$

In this particular case the Ward identity ($\hat{m} \equiv (m_u + m_d)/2$)

$$F_\pi M_\pi^2 = \hat{m} G_\pi \ , \tag{13}$$

which also holds in finite volume, makes the use of such a formula unnecessary: from the finite-volume version of eq. (13) one immediately obtains

$$\frac{\Delta G_\pi}{G_\pi} =: R_G = R_F + 2 R_M \ , \tag{14}$$

where R_M is the relative shift for M_π. On the other hand, since we have an explicit expression for all three relative shifts for large volumes, eq. (14) can be used as a nontrivial check on the asymptotic formulae. Indeed, all three relative shifts can be expressed as an integral with the same weight function, and eq. (14) can be satisfied only if the same relation holds among the integrands[2]:

$$\frac{\hat{m}}{M_\pi} N_G(\nu) = N_F(\nu) - \frac{F_\pi}{M_\pi} F(\nu) \ , \tag{15}$$

where $F(\nu) = T^{I=0}(0, \nu)$ is the forward scattering amplitude appearing in Lüscher's formula for M_π. It is easy to verify that this relation follows from the Ward identity[3]

$$-i Q^\mu \langle \pi^1(p_1) \pi^1(p_2) \pi^3(p_3) | A_\mu^3(0) | 0 \rangle = \hat{m} \langle \pi^1(p_1) \pi^1(p_2) \pi^3(p_3) | P^3(0) | 0 \rangle \ , \tag{16}$$

---

[2] We correct a typo in the published version. The prefactor $\hat{m}/M_\pi$ was omitted.
[3] Notice that in the definition of N_F, eqs. (5,6), the $\langle 3\pi | A_\mu | 0 \rangle$ amplitude is multiplied with p_3^μ and not with Q^μ as in this Ward identity.

once the limit to the relevant kinematical configuration is taken and if one properly accounts for the pole at $Q^2 = M_\pi^2$ present in both amplitudes.

5 The asymptotic formula in Chiral Perturbation Theory

As was shown in [2], the Lüscher formula for the pion mass can be used very conveniently in combination with the chiral expansion for the $\pi\pi$ scattering amplitude. The same can be done for F_π using the chiral expansion for the infinite-volume $\langle 3\pi | A_\mu | 0 \rangle$ amplitude, which has been calculated up to next-to-leading order in [6]. The chiral expansion for the amplitude N_F reads

$$N_F(\nu) = \frac{M_\pi}{F_\pi} \left[N_2^F(\tilde\nu) + \xi N_4^F(\tilde\nu) + O(\xi^2) \right] , \qquad (17)$$

where $\xi = (M_\pi/4\pi F_\pi)^2$ and $\tilde\nu = \nu/M_\pi$, and translates into a corresponding expansion for ΔF_π

$$R_F := \frac{\Delta F_\pi}{F_\pi} = \frac{6}{\lambda} \left[\xi I_2^F(\lambda) + \xi^2 I_4^F(\lambda) + O(\xi^3) \right] , \qquad (18)$$

where $\lambda = M_\pi L$. The integrals I_n can be given analytically in terms of a few basic integrals:

$$\begin{aligned}
I_2^F(\lambda) &= -2 B^0(\lambda) \qquad (19) \\
I_4^F(\lambda) &= \left(2\bar\ell_1 + \frac{4}{3}\bar\ell_2 - 3\bar\ell_4 - \frac{7}{9} \right) B^0(\lambda) + \left(-\frac{8}{3}\bar\ell_1 - \frac{32}{3}\bar\ell_2 + \frac{112}{9} \right) B^2(\lambda) \\
&+ \frac{4}{3} \left(R_0^0(\lambda) - R_0^1(\lambda) - 10 R_0^2(\lambda) \right) - \frac{13}{6} R_0^{0\prime}(\lambda) + \frac{8}{3} R_0^{1\prime}(\lambda) + \frac{20}{3} R_0^{2\prime}(\lambda) ,
\end{aligned}$$

where the integrals B^{2k} and R_i^k are defined as

$$B^{2k}(\lambda) = \int_{-\infty}^{\infty} d\tilde y \, \tilde y^{2k} \, e^{-\sqrt{1+\tilde y^2}\lambda} = \frac{\Gamma(k+1/2)}{\Gamma(3/2)} \left(\frac{2}{\lambda} \right)^k K_{k+1}(\lambda) , \qquad (20)$$

and

$$R_0^{k(l)}(\lambda) = \begin{cases} \mathrm{Re} \\ \mathrm{Im} \end{cases} \int_{-\infty}^{\infty} d\tilde y \, \tilde y^k \, e^{-\sqrt{1+\tilde y^2}\lambda} \, g^{(l)}(2(1+i\tilde y)) \quad \text{for} \begin{cases} k \text{ even} \\ k \text{ odd} \end{cases} , \qquad (21)$$

with[4]

---
[4] The function $g(x)$ is related to the standard $\bar J$ one-loop function through $g(x) = 16\pi^2 \bar J(x M_\pi^2)$.

$$g(x) = \sigma \log \frac{\sigma-1}{\sigma+1} + 2 \ , \qquad g'(x) = \frac{1}{x}\left[\frac{2}{\sigma x}\log\frac{\sigma-1}{\sigma+1} - 1\right] \ , \tag{22}$$

with $\sigma = \sqrt{1-4/x}$. These integrals (with the only exception of the primed R_0^k) have already been introduced in [2].

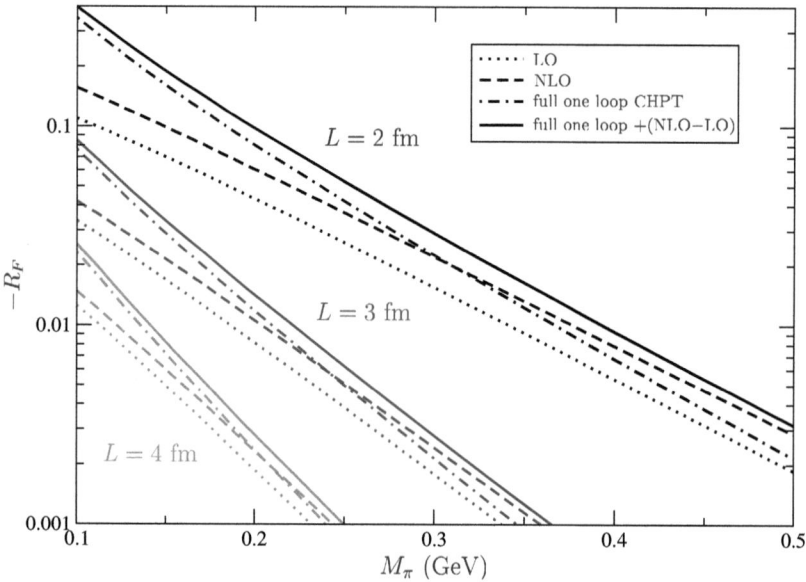

Fig. 2. The absolute value of the relative finite volume correction $R_F = F_{\pi,L}/F_\pi - 1$ as a function of M_π for different volume sizes. We plot the leading (LO) and next-to-leading order (NLO) in the chiral expansion of the asymptotic formula (18) and also the full one-loop result in CHPT (24). The solid lines show the sum of the full one-loop result and the NLO correction in the asymptotic formula.

We have evaluated numerically these corrections using the following values for the chiral low energy constants [7]:

$$\bar{\ell}_1 = -0.4 \pm 0.6, \qquad \bar{\ell}_2 = 4.3 \pm 0.1, \qquad \bar{\ell}_4 = 4.4 \pm 0.2 \ . \tag{23}$$

The results are displayed in fig. 2 where we plot the modulus of R_F as a function of M_π for volume sizes between 2 and 4 fm. We have studied the uncertainties in R_F which arise from the low energy constants (23) and found that they are barely visible on the plot – we therefore omit them (in size they are similar to the thickness of the lines). In the figure we compare the evaluation of the asymptotic formula to leading and next-to-leading order also to the full one-loop calculation of Gasser and Leutwyler [8], which can be given in a very compact form:

$$F_{\pi,L} = F_\pi \left[1 - \xi \tilde{g}_1(\lambda) + O(\xi^2)\right] \tag{24}$$

where

$$\tilde{g}_1(\lambda) = {\sum}' \int_0^\infty dx\, e^{-\frac{1}{x}-\frac{x}{4}(n_1^2+n_2^2+n_3^2)\lambda^2} , \tag{25}$$

where the prime indicates that the sum runs over all integer values of n_i, excluding the term with all $n_i = 0$.

In comparison to the pion mass, the finite volume corrections in eq. (24) are a factor 4 larger but negative – the sign difference is in accordance with the observation that in finite volume chiral symmetry is restored, i.e. the pion becomes heavier and its decay constant tends to vanish. Apart from this quantitative difference, the numerical analysis gives results which are qualitatively similar to those obtained for the pion mass [2]:

1. the finite volume corrections are exponentially suppressed for large values of $M_\pi L$ and become negligible rather quickly;

2. the leading term in the chiral expansion of the asymptotic formula receives large corrections even for the physical values of the quark masses – the similarity to the pion mass results makes us however think that the series will start to show a convergent behaviour at NNLO;

3. the leading term in the asymptotic expansion also receives large corrections from the subleading ones whenever the finite volume effects are nonnegligible;

4. since subleading terms are important both in the chiral as well as in the asymptotic expansion, the best estimate of the size of these finite-volume corrections is obtained by summing the subleading effects in both expansions, as shown by the solid curves in fig. 2.

For example, in a recent calculation of F_π on the lattice [9] with dynamical fermions a volume of $L = 2.5$ fm size has been used, and pion masses as low as 0.24 GeV. For these values we find that the finite volume corrections evaluated with the asymptotic formula to NLO (LO) are 1.5% (1.1%), whereas the full one-loop calculation gives 1.6%. Adding both types of subleading effects we find a total correction of 2%. In Ref. [10], $L = 1.5$ fm and $M_\pi = 0.4$ GeV were used: in this case the full one-loop calculation gives a 3.4% effect, whereas adding the NLO chiral corrections we get to 4.5%. For the parameters used in [11] finite-volume effects are negligible.

6 Conclusions

We have derived an asymptotic formula for the pion decay constant in a finite large volume along the same lines as Lüscher's formula for the pion mass [4]. The advantage offered by such a formula is a relatively easy access to a study of higher order chiral corrections in finite volume effects. We have evaluated these numerically and have shown that in F_π these corrections are large, analogously to what has been found for M_π [2]. In the present case we could use existing calculations of the relevant infinite-volume amplitude to evaluate next-to-leading chiral corrections. Going one order higher in this expansion would require the calculation of the $\langle 3\pi | A_\mu | 0 \rangle$ amplitude to two loops in CHPT.

The asymptotic formula derived here immediately applies (after the necessary but obvious modifications) to other similar quantities, like G_π. As we have explicitly verified, the asymptotic formulae for F_π and G_π satisfy a Ward identity that relates their ratio to $M_\pi^2/\hat{m}$: if one extracts the finite-volume expression for M_π from this Ward identity one recovers Lüscher's formula. The formula applies also to the decay constants of heavier mesons, like F_K. In the latter case the study of these finite volume effects [12] is of direct phenomenological interest in view of the recent application of the lattice calculation of the F_K/F_π ratio to the extraction of V_{us} [13] – it is worth mentioning that for this application the required precision of the lattice result is at the percent level. The same formula can also be applied to the decay constants of yet heavier mesons, like f_D or f_B. In this case, however, the advantage provided by the asymptotic formula with respect to a plain one-loop calculation (as recently performed in [14]) will be of practical relevance only if the knowledge of the low energy constants of the chiral Lagrangian describing the coupling of heavy mesons to pions [15] is extended beyond leading order.

Acknowledgments

We thank Stephan Dürr, Heiri Leutwyler, Martin Lüscher and Rainer Sommer for useful discussions and/or comments on the manuscript. This work is supported by the Swiss National Science Foundation and in part by RTN, BBW-Contract No. 01.0357 and EC-Contract HPRN–CT2002–00311 (EURIDICE).

References

[1] J. Gasser and H. Leutwyler, Nucl. Phys. B 307 (1988) 763.

[2] G. Colangelo and S. Durr, Eur. Phys. J. C 33 (2004) 543 [hep-lat/0311023].

[3] D. Becirevic and G. Villadoro, Phys. Rev. D 69 (2004) 054010 [hep-lat/0311028];
A. Ali Khan et al. [QCDSF-UKQCD Coll.], [hep-lat/0312030];
M. Guagnelli et al., [hep-lat/0403009];
D. Arndt and C. J. D. Lin, [hep-lat/0403012];
S. R. Beane, [hep-lat/0403015].

[4] M. Lüscher, Commun. Math. Phys. 104, 177 (1986).

[5] J. Bijnens et al., Phys. Lett. B 374 210 (1996) [hep-ph/9511397],
Nucl. Phys. B 508 263 (1997) [Erratum-ibid. B 517 639 (1998)] [hep-ph/9707291].

[6] G. Colangelo, M. Finkemeier and R. Urech, Phys. Rev. D 54 (1996) 4403 [hep-ph/9604279].

[7] G. Colangelo, J. Gasser and H. Leutwyler, Nucl. Phys. B 603, 125 (2001) [hep-ph/ 0103088].

[8] J. Gasser and H. Leutwyler, Phys. Lett. B 184 83 (1987).

[9] C. T. H. Davies et al. [HPQCD Coll.], Phys. Rev. Lett. 92 (2004) 022001 [hep-lat/0304004];
C. Aubin et al. [MILC Coll.], [hep-lat/0309088].

[10] C. R. Allton et al. [UKQCD Coll.], [hep-lat/0403007].

[11] F. Farchioni, I. Montvay and E. Scholz [qq+q Coll.], [hep-lat/0403014].

[12] G. Colangelo and C. Haefeli, in preparation.

[13] W. J. Marciano, [hep-ph/0402299].

[14] D. Arndt and C. J. D. Lin, in Ref. [3]

[15] G. Burdman and J. F. Donoghue, Phys. Lett. B 280, 287 (1992),
M. B. Wise, Phys. Rev. D 45 (1992) 2188.
T. M. Yan et al., Phys. Rev. D 46 (1992) 1148 [Erratum-ibid. D 55 (1997) 5851].

II

Finite volume effects for meson masses and decay constants

published in

Nucl. Phys. B721 (2005) 136

Finite volume effects for meson masses and decay constants

Gilberto Colangelo, Stephan Dürr and Christoph Haefeli

Institut für Theoretische Physik, Universität Bern
Sidlerstr. 5, 3012 Bern, Switzerland

Abstract

We present a detailed numerical study of finite volume effects for masses and decay constants of the octet of pseudoscalar mesons. For this analysis we use chiral perturbation theory and asymptotic formulae à la Lüscher and propose an extension of the latter beyond the leading exponential term. We argue that such a formula, which is exact at the one-loop level, gives the numerically dominant part at two loops and beyond. Finally, we discuss the possibility to determine low energy constants from the finite volume dependence of masses and decay constants.

1 Introduction

In lattice QCD the determination of the mass and decay constant of the lowest-lying state with a given set of quantum numbers is entering the high-precision era. Even in the fully unquenched case (i.e. with sea- and valence-quarks being degenerate) the pion mass can be measured, for fixed bare parameters, with an accuracy at the percent level, and future progress towards the permille level is anticipated. However, to make contact with the real world, three extrapolations are needed. These are (i) the continuum extrapolation, (ii) the infinite volume extrapolation, and (iii) the chiral extrapolation. In each case there is considerable help and an analytical guideline from an effective field theory framework.

This article is concerned with the extrapolation to infinite volume, where the situation is particularly favorable (for a compact review of the recent literature see [1]). As shown by Gasser and Leutwyler, chiral symmetry imposes strong constraints on the dynamics at low energy in QCD, even if the system is enclosed in a finite box. Accordingly, Chiral Perturbation Theory (ChPT) may be adapted to the finite volume case [2, 3, 4]. In this framework finite volume effects can be taken systematically into account, in a perturbative loop expansion. If the spatial volume L^3 is large enough internal degrees of freedom of the particle of interest play no role as far as finite volume effects are concerned and these are exclusively due to pion loops. This means that they first appear at next-to-leading order (NLO), i.e. at $O(p^4)$ in the chiral counting. Another consequence is that they are exponentially small in the pion mass for any particle that couples to the pion field: the effect behaves like $e^{-M_\pi L}$ for pions, kaons, etas and nuclei. To date, a number of finite volume calculations have been performed at one-loop order. In the nucleon sector the corrections to the mass [5, 6], the magnetic moment [6] and the baryon axial charge [7] have been worked out. In the meson sector the original calculation for $M_\pi(L)$ and $F_\pi(L)$ [2] has been extended to the quenched case [8]. The same quantities have also been analyzed in a quark-meson model [9]. More recently, the finite volume shifts for $F_K(L)$ and $B_K(L)$ have been given in Ref. [10] and the extension to heavy-meson chiral perturbation theory has been described [11]. No full two-loop calculation of finite volume effects has appeared yet.

For some observables the Lüscher formula represents a convenient and powerful alternative [12]. It allows to estimate subleading (in the chiral counting) finite volume effects for the mass of a particle with less effort, while sticking to the leading order in an expansion in powers of $e^{-M_\pi L}$. The formula gives the finite volume shift $M_P(L)-M_P$ of a particle P in terms of the infinite volume πP forward scattering amplitude in the unphysical (Euclidean) region. For this amplitude the ChPT expression at a certain loop order is used. The approach via the Lüscher formula is economical, since only a chiral calculation in infinite volume is needed, and the loop in finite volume comes for free. Applications of the Lüscher formula to the mass of the nucleon [5, 13] and the pion [14] have been worked out. In the latter publication several orders in the ChPT input for the $\pi\pi$ scattering amplitude were compared and it was found that in a certain range of M_π and L subleading effects (in the chiral counting) can be large with respect to the leading contributions. An extension of the Lüscher formula has been constructed for pseudoscalar decay constants [15].

This article presents a resummed version of the Lüscher formulae for masses and decay constants (this has been briefly discussed by one of us in [1]), where the terms neglected in the large-L expansion are $\sim e^{-(\sqrt{3}+1)/\sqrt{2}\cdot M_\pi L}$, rather than $e^{-\sqrt{2}M_\pi L}$. We proceed with the asymptotic expression for $M_\pi(L) - M_\pi$ to 3-loop order, and for $F_\pi(L)-F_\pi, M_K(L)-M_K, F_K(L)-F_K, M_\eta(L)-M_\eta$ to 2-loop order, by using the available knowledge in the literature on the scattering amplitude or axial-vector matrix element. In all cases, the result may be given in a compact formula that does not involve any numerical integration.

2 Finite volume effects

Lattice calculations are necessarily done in a finite 4D volume which acts as an IR-cutoff. Typically a $L^3 \times T$ geometry with periodic boundary conditions in all directions is chosen with $T \gg L$ and both large compared to the inverse temperature of the QCD phase transition or crossover. In order to determine the mass of a particle one considers a correlator of two properly chosen interpolating fields

$$C(t) = \int d^3x \, \langle \phi(x)\phi(0) \rangle \, e^{\mathrm{i}\mathbf{p}\mathbf{x}} \stackrel{T \to \infty}{\longrightarrow} \sum_{n=0}^{\infty} c_n \, e^{-E_n t} \qquad (1)$$

and tries to determine the energy levels (typically the lowest) at zero spatial momentum, $M_n(L) = E_n(\mathbf{p}=0)$. In the following, we will be concerned with the shift $M(L) - M$, where $M \equiv M(L=\infty)$, for the groundstate ($n=0$) due to the finite 3D volume L^3. Decay constants or, more generally, matrix elements suffer from analogous shifts at finite spatial volume. For not-too-small box length L these shifts can be calculated analytically, thus offering a means to correct lattice data for this systematic effect. We now give a brief outline of the two main frameworks for such a calculation. A comment on cut-off effects and their interplay with finite volume effects is given in app. B.

2.1 ChPT in finite volume

In QCD with light flavors the physics in the infrared region is controlled by chiral symmetry. As shown by Gasser and Leutwyler this still holds true if the system is enclosed in a finite box $L^3 \times T$, provided both L and T are large enough that chiral symmetry is not restored [4]. They have shown that in an isotropic box with periodic boundary conditions for the meson fields the finite-volume dependence comes in exclusively through the propagators. The latter becomes periodic in all spatial directions, and in the limit $T \to \infty$ can be written as follows

$$G(x^0, \mathrm{x}) = \sum_n G_0(x^0, \mathrm{x} + \mathrm{n}L) \tag{2}$$

which is equivalent to replacing the integration over the spatial part of the momenta by a sum over multiples of $2\pi/L$. In other words, with periodic boundary conditions the Lagrangian remains the same as in infinite volume.

The expansion parameters in ChPT are

$$\frac{p}{4\pi F_\pi} \quad , \quad \frac{M_\pi}{4\pi F_\pi} \tag{3}$$

and the theory can be meaningfully applied only if both are small. In a finite volume spatial momenta are discretized: $p = 2\pi n/L$ with n a vector of integers. Therefore one can have "small" nonzero momenta and apply ChPT only if the condition

$$L \gg \frac{1}{2F_\pi} \sim 1\,\mathrm{fm} \tag{4}$$

is satisfied. A priori there is no way to say how much L has to be in excess of 1 fm. As a guideline we observe that the lowest non-trivial momentum in a 1 fm box is 1.2 GeV, which is certainly beyond the realm of ChPT. Note, finally, that unlike $F_\pi L$ the combination $M_\pi L$ is not constrained. Both $M_\pi L \ll 1$ and $M_\pi L \gg 1$ are acceptable [2, 3, 4, 16], but they imply different ways to organize the chiral series,

$$M_\pi L \gg 1 \quad \leftrightarrow \quad \text{``p-expansion''} \tag{5}$$
$$M_\pi L \ll 1 \quad \leftrightarrow \quad \text{``ϵ-expansion''}. \tag{6}$$

Here we shall restrict ourselves to the former case, where the chiral counting is

$$M_\pi^2 \sim m \sim O(p^2), \qquad 1/L \sim p \sim O(p). \tag{7}$$

With this setup Gasser and Leutwyler calculated the mass and decay constant shift in a theory with $N_f^2 - 1$ degenerate pseudo-Goldstone bosons, and obtained [2]

$$M_\pi(L) = M_\pi \left[1 + \frac{1}{2N_f} \xi_\pi \tilde{g}_1(\lambda_\pi) + O(\xi_\pi^2) \right] \tag{8}$$

$$F_\pi(L) = F_\pi \left[1 - \frac{N_f}{2} \xi_\pi \tilde{g}_1(\lambda_\pi) + O(\xi_\pi^2) \right]. \tag{9}$$

n	1	2	3	4	5	6	7	8	9	10
$m(n)$	6	12	8	6	24	24	0	12	30	24
n	11	12	13	14	15	16	17	18	19	20
$m(n)$	24	8	24	48	0	6	48	36	24	24

Tab. 1. The multiplicities $m(n)$ in (12) for $n \leq 20$.

Here we have introduced the abbreviations (note the F_π in the denominator for all P)

$$\xi_P \equiv \frac{M_P^2}{(4\pi F_\pi)^2} \qquad (10)$$

$$\lambda_P \equiv M_P L \qquad (11)$$

for $P=\pi$ (and $P=K, \eta$ will be used below) as well as the modified [1] shape function

$$\tilde{g}_1(x) = \sum_{n=1}^{\infty} \frac{4m(n)}{\sqrt{n}\,x} K_1(\sqrt{n}\,x) \qquad (12)$$

where K_1 is a Bessel function of the second kind and the multiplicities $m(n)$ have been given in [14], but for convenience we reproduce them in tab. 1. Given the asymptotic expansion $K_1(z) \sim \sqrt{\pi/(2z)}\,e^{-z}$, it is clear that in the p-regime eqns. (8, 9, 12) represent quickly converging expressions. Several observables have been worked out at one-loop order [5, 6, 7, 8, 10, 11, 17, 18], but to date no two-loop result obtained in this setup has appeared.

2.2 Lüscher formula

An entirely independent approach has been devised by Lüscher who has proven an elegant relation between the mass shift of the particle P in a finite volume $L^3 \times \infty$ and the $P\pi$ scattering amplitude in infinite volume [12]

$$M_P(L) - M_P = -\frac{3}{16\pi^2 \lambda_P} \int_{-\infty}^{\infty} dy\, \mathcal{F}_P(iy)\, e^{-\sqrt{M_\pi^2+y^2}\,L} + O(e^{-\bar{M}L})\,. \qquad (13)$$

Here $\mathcal{F}_P(\nu)$ denotes the infinite volume forward ($t=0$) scattering amplitude of P and π in Minkowski space. The integration runs along the imaginary axis, i.e. $\mathcal{F}_P(\nu)$ is evaluated for

---
[1] Our $\tilde{g}_1$ relates to g_1 of [2] via $\tilde{g}_1(\lambda_\pi) = (4\pi/M_\pi)^2 \cdot g_1(M_\pi, \beta=\infty, L)$ and is a dimensionless function.

$$\nu = iy \tag{14}$$

with real y, thus staying far away from the cuts. Only the real part of $\mathcal{F}_P(iy)$ contributes to the integral, since the imaginary part is odd in y. An additional piece in the original formula, referring to the 3-particle vertex, is omitted here, since we assume P to be a pseudo-Goldstone boson. The Lüscher formula (13) keeps only the leading term in an expansion in (fractional) powers of $e^{-\lambda_\pi}$. The generic bound $\bar{M} \geq \sqrt{3/2} M_\pi$ can be specified to $\bar{M} = \sqrt{2} M_\pi$ in a theory with pseudo-Goldstone bosons only.

Recently, an analogous "Lüscher-type" formula has been derived for the finite volume shift of the axial-vector decay constant F_P. It reads [15]

$$F_P(L) - F_P = +\frac{3}{8\pi^2 \lambda_P} \int_{-\infty}^{\infty} dy\, \mathcal{N}_P(iy)\, e^{-\sqrt{M_\pi^2 + y^2} L} + O(e^{-\bar{M} L}) \tag{15}$$

where $\mathcal{N}_P(\nu)$ is derived from the matrix element $\langle \pi\pi | A_\mu | P \rangle$ via a subtraction prescription we will specify below. Like in the mass formula (13) the finite volume shift of F_P is expressed in terms of an infinite-volume amplitude, evaluated in the unphysical (Euclidean) region, thus far away from the cuts. Again, only the leading term in an expansion in (fractional) powers of $e^{-\lambda_\pi}$ is kept. Note the reverse overall sign, compared to (13), and the fact that the net physical effect is opposite to this prefactor; in other words $M_P(L) > M_P$ and $F_P(L) < F_P$.

To predict the shifts $M_P(L) - M_P$ and $F_P(L) - F_P$ in a lattice calculation with a known box length L, and thus to correct the data for this systematic effect, the formulae (13, 15) must be fed with an explicit representation of the amplitudes $\mathcal{F}_P(\nu)$ and $\mathcal{N}_P(\nu)$, respectively. This is the place where ChPT naturally enters, even if one opts for the Lüscher approach. Using existing knowledge about the relevant amplitude at n-loop order, one gets the leading piece, in the $e^{-\lambda_\pi}$ expansion, of the finite-volume shift of $M_P(L), F_P(L)$ to $n{+}1$-loop order. For instance, using the tree-level expressions $\mathcal{F}_\pi(\nu) = -M_\pi^2/F_\pi^2$ and $\mathcal{N}_\pi(\nu) = -2M_\pi/F_\pi$ in 2-flavor ChPT yields

$$M_\pi(L) - M_\pi = +\frac{3}{8\pi^2} \frac{M_\pi^2}{F_\pi^2 L} K_1(\lambda_\pi) + O(e^{-\sqrt{2}\lambda_\pi}) \tag{16}$$

$$F_\pi(L) - F_\pi = -\frac{3}{2\pi^2} \frac{M_\pi}{F_\pi L} K_1(\lambda_\pi) + O(e^{-\sqrt{2}\lambda_\pi}) \tag{17}$$

in agreement [2] with the 1-loop chiral expressions (8) and (9). Because of the "elevator"-effect in the loop expansion and since the associated chiral calculation is in infinite volume, it is much easier to push to higher chiral orders in the Lüscher-type setup than with a straightforward ChPT-in-finite-volume calculation [19]. This offers a genuine opportunity to compare several chiral orders and thus to assess the chiral convergence behavior. Indeed, in [14] the finite volume shift in the pion mass was evaluated, using ChPT at LO/NLO/NNLO for $\mathcal{F}_\pi(\nu)$ to get the asymptotic piece of the full chiral expression at 1/2/3-loop order, and it was found that for some (M_π, L)-combinations the chiral series converges well, if at least the NLO input is included. Still, one might worry whether the non-asymptotic pieces of order $O(e^{-\sqrt{2}\lambda_\pi})$ would prove numerically relevant [19]. In [14] a first attempt was made to discriminate those regions in the (M_π, L)-plane where higher orders in the chiral expansion dominate against those regions where terms omitted in the Lüscher approach are more important. Below, we shall present a resummed version of the Lüscher-type formulae (13, 15), where the pieces $\propto e^{-\sqrt{2}\lambda_\pi}$ and $\propto e^{-\sqrt{3}\lambda_\pi}$ are included and the terms $O(e^{-(\sqrt{3}+1)/\sqrt{2}\cdot\lambda_\pi})$ estimated. On this basis a more precise assessment of the relevance of higher loop corrections versus higher powers of $e^{-\lambda_\pi}$ can be made.

Note finally that all Lüscher-type formulae build on the unitarity of the theory and thus hold for the full (unquenched) theory. In the (partially) quenched case it seems indispensable to start in the framework of subsect. 2.1, but even then the arguments for using the infinite volume Lagrangian reside on less solid grounds – see Ref. [8] for a lucid discussion.

3 The Lüscher formula resummed

In Lüscher's derivation of the asymptotic formula for the finite volume correction to particle masses the first step is a proof that the leading exponential term is given by the sum of all diagrams in which only one propagator is taken in finite volume. This class of diagrams yields

$$M_P(L) - M_P = -\frac{1}{4M_P} \sum_{n \neq 0} \int \frac{d^4q}{(2\pi)^4} e^{iq\cdot nL} G_0(q) \Gamma(\hat{p}, q, -\hat{p}, -q) + \ldots \quad (18)$$

where $G_0(q) = 1/(M_\pi^2 + q^2)$ is the full propagator and $\Gamma(p_1, p_2, p_3, p_4)$ the four-point vertex function in infinite volume. Lüscher then concentrates on the leading exponential contributions (those with $|n| = 1$), and shows that, if one disregards terms which are exponentially suppressed with respect to $\exp(-\lambda_\pi)$, three of the four integrations in (18) can be performed explicitly and the result (13) is obtained. The same reasoning, however, applies also to all other terms in the sum in (18): for each of the terms with $|n| > 1$, one can obtain its leading exponential contribution by performing exactly the same steps that Lüscher did for the $|n| = 1$ term and work out three of the four integrations explicitly. It is easy to keep track of the vector n in doing these manipulations, and to get the resummed formula

$$M_P(L) - M_P = -\frac{1}{32\pi^2 \lambda_P} \sum_{n=1}^{\infty} \frac{m(n)}{\sqrt{n}} \int_{-\infty}^{\infty} dy \, \mathcal{F}_P(iy) \, e^{-\sqrt{n(M_\pi^2 + y^2)}L} + O(e^{-\bar{M}L}) \quad (19)$$

where $m(n)$ has been given in tab. 1 and $\mathcal{F}_P(\nu)$ is the $P\pi$ forward scattering amplitude as usual. The extension which we are proposing is done in the same spirit as the extension of the domain of integration in (19) to infinity (the contributions from the region $|y| > \sqrt{\bar{M}^2 - M_\pi^2}$ are beyond the accuracy of the formula): being of "kinematical" nature, the extension comes at no cost and may be numerically relevant. In this case, actually, one even obtains an improvement in the algebraic accuracy of the formula: we now have $\bar{M} > \sqrt{3}M_\pi$ and not, as before, $\bar{M} = \sqrt{2}M_\pi$. In other words, the terms $O(e^{-\sqrt{2}\lambda_\pi})$ and $O(e^{-\sqrt{3}\lambda_\pi})$ are included now. To further clarify the meaning of (19) let us list the two main classes of exponentially suppressed contributions which are still missing:

1. All diagrams which have more than one pion loop in finite volume. Obviously, these contributions start at the two-loop level in the chiral expansion.

2. Contributions to the integral (18) which are due to singularities in either the propagator G_0 or the vertex function Γ which are further away from the real axis than M_π. These singularities show up only if one considers the vertex function at one loop, or the propagator at two-loop accuracy and beyond. All these contributions appear only if one calculates finite-volume effects at the two-loop level in the chiral expansion.

In the first class we distinguish those diagrams where different loops factorize (i.e. loops which have no propagator in common) from those which do not. It is easy to see that the former sub-class yields corrections of order $e^{-2\lambda_\pi}$. For two-loop diagrams which do not factorize (e.g. the two-loop sunset diagram) on may rely on the general discussion by Lüscher (cf. in particular Eqn. (2.49) in Ref. [12]) and conclude that the sunset diagram decays exponentially at large L as $\exp(-N_{\text{sunset}}\lambda_\pi)$ with $N_{\text{sunset}} = (\sqrt{3}+1)/\sqrt{2} \simeq 1.93$. For the second class we remark that the singularities neglected in (19) are due to the exchange of at least two pions, hence starting at $\nu = \pm 2M_\pi$, and this gives terms of order $e^{-2\lambda_\pi}$. We have not tried to prove the statement that the algebraic accuracy of our formula is given by

$$\bar{M} = M_\pi(\sqrt{3}+1)/\sqrt{2} \qquad (20)$$

beyond the two-loop level, since it appears to us to be a question of academic interest.

More interesting is the question of how accurate the formula (19) is numerically. A complete analysis of finite-volume effects for M_π and F_π at the two loop level, which is currently under way [20], will clarify this point. The partial results we have so far indicate that the formula is very accurate. We have tried to find an algebraic reason for this, and found out that at the two loop level all diagrams which do not appear in eq. (19) are suppressed (besides an extra exponential factor) by some power of $1/L$. The numerical results, however, seem to go beyond what one would expect from such an argument.

An analogy to the low-temperature expansion seems more suggestive. In the effective theory the large volume and the low temperature expansions are in one-to-one correspondence [4]. In Ref. [21] Schenk discussed the propagation of pions through matter in a state of thermal equilibrium at inverse temperature β. If the temperature is not too high, the hadronic phase mainly consists of pions, with effects of other excitations such as K, η, ρ... exponentially suppressed. Due to interactions with pions of the heat bath, the effective pion mass $M_\pi(\beta)$ is given by [21]

$$M_\pi(\beta) - M_\pi = -\frac{1}{2M_\pi} \int \frac{d^3q}{(2\pi)^3 2\omega_q} \, n_B(\omega_q) \, T_{\pi\pi}^{I=0}(s) + O(n_B^2) \qquad (21)$$

with $\omega_q = \sqrt{M_\pi^2 + \mathbf{q}^2}$ and the density $n_B(x) = 1/(e^{\beta x} - 1)$. The details of the pion kinematics will be discussed in the next section, and the isospin index refers to the t-channel. One immediately verifies that (21) agrees with the modified Lüscher formula in one dimension to first order in the density n_B – the density factor n_B in the integrand is the outcome of the resummation over n in (18) if n is taken as a one-dimensional vector. Schenk has carried the expansion of (21) one step further and determined the contributions of order n_B^2 to the pion mass at finite temperature: in this extension, effects generated by three-body collisions are explicitly accounted for. It turned out that these effects are numerically very small, in line with intuition – the rescattering of three pions into three pions is a rare process unless the density is very high. Although the argument cannot be formulated in the same way for the finite volume case, we do see that the outcome of the numerical analysis is the same.

4 Meson masses and decay constants in finite volume

We start with the resummed Lüscher formulae for the relative finite-size shift of pseudoscalar masses and decay constants

$$R_{M_P} \equiv \frac{M_P(L) - M_P}{M_P}$$

$$= -\frac{M_\pi}{32\pi^2 M_P \lambda_P} \sum_{n=1}^{\infty} \frac{m(n)}{\sqrt{n}} \int_{-\infty}^{\infty} d\tilde{y}\, \mathcal{F}_P(i\tilde{y}) e^{-\sqrt{n(1+\tilde{y}^2)}\lambda_\pi} + O(e^{-\bar{M}L}) \quad (22)$$

$$R_{F_P} \equiv \frac{F_P(L) - F_P}{F_P}$$

$$= +\frac{M_\pi}{16\pi^2 F_P \lambda_P} \sum_{n=1}^{\infty} \frac{m(n)}{\sqrt{n}} \int_{-\infty}^{\infty} d\tilde{y}\, \mathcal{N}_P(i\tilde{y}) e^{-\sqrt{n(1+\tilde{y}^2)}\lambda_\pi} + O(e^{-\bar{M}L}) \quad (23)$$

where all symbols on the r.h.s. refer to infinite volume quantities. The multiplicities $m(n)$ are given in tab. 1, λ_P has been defined in (11), and the dimensionless integration variable (which we will be using from here on) relates to the previous one via $\tilde{y} = y/M_\pi$. The amplitudes $\mathcal{F}_P(\tilde{\nu}), \mathcal{N}_P(\tilde{\nu})$ with $\tilde{\nu} = \nu/M_\pi$ are

$$\mathcal{F}_P(\tilde{\nu}) = T_{\pi P}^{I=0}(0, -4M_P \nu)$$

$$\mathcal{N}_P(\tilde{\nu}) = -i\bar{A}_P^{I=0}(0, -4M_P \nu) \quad (24)$$

where $T_{\pi P}^{I=0}(t, u-s)$ is the $P\pi$-scattering amplitude with zero t-channel isospin, and $\bar{A}_P^{I=0}(t, u-s)$ is the subtracted amplitude for the decay of the meson P with momentum p into two pions in an isospin zero state (in the t-channel) via an axial current insertion. The subtraction removes the one-particle reducible contribution and is defined through

$$\bar{A}_P^{I=0}(t, u - s) = \frac{p^\mu}{M_P}(\bar{A}_P^{I=0})_\mu$$

$$(\bar{A}_P^{I=0})_\mu = (A_P^{I=0})_\mu - iQ_\mu F_P \frac{T_{\pi P}^{I=0}(t, u-s)}{Q^2 - M_P^2}$$

$$(A_P^{I=0})_\mu = \langle (\pi(p_1)\pi(p_2))_{I=0} | A_\mu(0) | P(p) \rangle \quad (25)$$

with $Q = p - p_1 - p_2$. Notice that the axial current in (25) must be normalized such that $\langle 0|A_\mu(0)|P(p)\rangle = ip_\mu F_P$. The amplitudes $\mathcal{F}_P(\tilde{\nu}), \mathcal{N}_P(\tilde{\nu})$ for $P = \pi, K, \eta$ have all (with the exception of $\mathcal{N}_\eta(\tilde{\nu})$, see below) been calculated at least at the one-loop level in ChPT. They have the generic form

$$\mathcal{F}_P(\tilde{\nu}) = \mathcal{F}_P^{(2)}(\tilde{\nu}) + \xi_P \mathcal{F}_P^{(4)}(\tilde{\nu}) + \xi_P^2 \mathcal{F}_P^{(6)}(\tilde{\nu}) + \mathcal{O}(\xi_P^3)$$

$$\mathcal{N}_P(\tilde{\nu}) = \mathcal{N}_P^{(2)}(\tilde{\nu}) + \xi_P \mathcal{N}_P^{(4)}(\tilde{\nu}) + \xi_P^2 \mathcal{N}_P^{(6)}(\tilde{\nu}) + \mathcal{O}(\xi_P^3)$$

with ξ_P defined in (10). Inserting such an expansion of the amplitude in (22, 23) leads to

$$R_{M_P} = -\sum_{n=1}^\infty \frac{m(n)}{2\sqrt{n}} \frac{1}{\lambda_\pi} \frac{M_\pi}{M_P} \xi_P \left[I_{M_P}^{(2)} + \xi_P I_{M_P}^{(4)} + \xi_P^2 I_{M_P}^{(6)} + \mathcal{O}(\xi_P^3) \right] \tag{26}$$

$$R_{F_P} = +\sum_{n=1}^\infty \frac{m(n)}{\sqrt{n}} \frac{1}{\lambda_\pi} \frac{F_\pi}{F_P} \xi_\pi \left[I_{F_P}^{(2)} + \xi_P I_{F_P}^{(4)} + \xi_P^2 I_{F_P}^{(6)} + \mathcal{O}(\xi_P^3) \right] \tag{27}$$

where the $I_{M_P}^{(2/4/6)}, I_{F_P}^{(2/4/6)}$ can be written in terms of a few basic integrals, as reported in Sect. 5. Please note the relative factor 2 in these two equations, to be consistent with [14, 15]. In the following we elaborate on the explicit form of (24) for the pion, kaon and eta.

4.1 Pion

The amplitudes $\mathcal{F}_\pi(\tilde{\nu}), \mathcal{N}_\pi(\tilde{\nu})$ defined in eq. (24) have been given in Refs. [14, 15], respectively. We give them here for convenience.

4.1.1 Pion mass

Consider (Minkowski space) $\pi\pi$-scattering

$$\pi(p_1) + \pi(p_2) \to \pi(p_3) + \pi(p_4) \tag{28}$$

with the kinematics

$$s = (p_1 + p_2)^2, \qquad t = (p_1 - p_3)^2, \qquad u = (p_1 - p_4)^2. \tag{29}$$

Isospin decomposition allows one to construct the t-channel isospin zero amplitude

$$T_{\pi\pi}^{I=0}(t, u-s) = A(s, t, u) + 3A(t, s, u) + A(u, s, t) \tag{30}$$

from the invariant amplitude $A(s,t,u)$ [22]. The amplitude entering the Lüscher formula follows by imposing the forward scattering kinematics $t=0$, viz.

$$\mathcal{F}_\pi(\tilde{\nu}) = T_{\pi\pi}^{I=0}(0, -4M_\pi\nu) \ . \tag{31}$$

4.1.2 Pion decay constant

We adopt the notation of Ref. [15], which relates to the one used in Eq. (25) by means of the one pion in the initial state being transferred to an outgoing pion. Crossing symmetry relates the two via $p_3 = -p$. The amplitude for the creation of three pions out of the vacuum with an axial current has been performed up to NLO in Ref. [23]. It is decomposed according to

$$\langle \pi^1(p_1)\pi^1(p_2)\pi^3(p_3) | A_\mu^3(0) | 0 \rangle = (p_1 + p_2)_\mu\, G + (p_1 - p_2)_\mu\, H + p_{3\mu}\, F \tag{32}$$

with the three scalar functions [2] $F = F(s_1, s_2, s_3)$, $G = G(s_1, s_2, s_3)$ and $H = H(s_1, s_2, s_3)$. The superscripts on the pion states and axial current are isospin indices. We have employed the variables $s_1 = (p_2 + p_3)^2$ and cyclic permutations. The combination which has two of the outcoming pions in an $I = 0$ state in the s_3-channel is given by

$$\langle (\pi(p_1)\pi(p_2))_{I=0}\pi^3(p_3) | A_\mu^3(0) | 0 \rangle = (A_\pi^{I=0})_\mu \tag{33}$$

$$\begin{aligned}(A_\pi^{I=0})_\mu &= (p_1 + p_2)_\mu\, G_0(s_1, s_2, s_3) + (p_1 - p_2)_\mu\, H_0(s_1, s_2, s_3) \\ &\quad + (p_3)_\mu\, F_0(s_1, s_2, s_3)\end{aligned}$$

with the isospin projected [3] form factors

$$F_0(s_1, s_2, s_3) = 3F_{123} + G_{231} + G_{312} - H_{231} + H_{312}$$

$$G_0(s_1, s_2, s_3) = 3G_{123} + \frac{1}{2}[F_{231} + F_{312} + G_{231} + G_{312} + H_{231} - H_{312}]$$

$$H_0(s_1, s_2, s_3) = 3H_{123} + \frac{1}{2}[F_{231} - F_{312} - G_{231} + G_{312} - H_{231} - H_{312}]$$

where $X_{ijk} = X(s_i, s_j, s_k)$ and $X = F, G, H$. The pole that the amplitude $(A_\pi^{I=0})_\mu$

---

[2] Note that our functions F, G, H are $1/\sqrt{2}$ times those of Ref. [23], in agreement with the notation in Ref. [15].

[3] The relation for G_0 is given for completeness; on imposing forward scattering kinematics, it will drop out.

has in the unphysical region $Q^2 = M_\pi^2$, $Q = -(p_1+p_2+p_3)$, needs to be subtracted as specified in (25)

$$(\bar{A}_\pi^{I=0})_\mu = (A_\pi^{I=0})_\mu - \mathrm{i}Q_\mu F_\pi \frac{T_{\pi\pi}^{I=0}(s_3, s_1 - s_2)}{Q^2 - M_\pi^2} \tag{34}$$

and in the end the result is evaluated in the forward kinematic configuration, i.e. for $s_3 = 0$. Hence the function $\bar{A}_\pi^{I=0}$ is a function of just one variable $\nu = (s_2 - s_1)/(4M_\pi)$, viz.

$$\bar{A}_\pi^{I=0}(0, -4M_\pi\nu) = 2\nu h_0(\nu) + M_\pi \bar{f}_0(\nu) \tag{35}$$

$$h_0(\nu) = H_0(2M_\pi(M_\pi - \nu), 2M_\pi(M_\pi + \nu), 0)$$

$$\bar{f}_0(\nu) = \bar{F}_0(2M_\pi(M_\pi - \nu), 2M_\pi(M_\pi + \nu), 0)$$

where the bar on the F_0 form factor indicates that it is defined after the subtraction of the pion pole (H_0 remains unaffected). The amplitude which enters the formula for $F_\pi(L)$ is then

$$\mathcal{N}_\pi(\tilde{\nu}) = -\mathrm{i}\,\bar{A}_\pi^{I=0}(0, -4M_\pi\nu) \,. \tag{36}$$

4.2 Kaon

The amplitudes $\mathcal{F}_K(\tilde{\nu})$ and $\mathcal{N}_K(\tilde{\nu})$ may be extracted from the form factors of the K_{l4} decay, as calculated in Ref. [24] up to NLO. Furthermore, an explicit representation of $\mathcal{F}_K(\tilde{\nu})$ is found in the πK-scattering study of Ref. [25] which has been extended to NNLO in Ref. [26]. Below, we stick to the NLO input, both for $\mathcal{F}_K(\tilde{\nu})$ and $\mathcal{N}_K(\tilde{\nu})$. For the mass the finite volume corrections at 2-loop level are very small (see discussion below) and a NNLO input is therefore not of particular interest for a lattice application. For the decay constant the result of the Lüscher formula with NLO input is not particularly small (see below), and a NNLO refinement would be useful. However, in this case one of the ingredients (to be precise: one of the 2-loop form factors of the axial-vector matrix element discussed in [27]) is missing. In the following we establish the relation of the amplitudes $\mathcal{F}_K(\tilde{\nu})$ and $\mathcal{N}_K(\tilde{\nu})$ to results given in the literature.

4.2.1 Kaon mass

The πK-scattering amplitude $T_{\pi K}(s,t,u)$ has been calculated at NLO in Ref. [25][4]

$$\pi^+(p_1) + K^+(p_2) \to \pi^+(p_3) + K^+(p_4) \qquad (37)$$

with the Mandelstam variables (29). This gives the t-channel isospin zero amplitude via

$$T_{\pi K}^{I=0}(t, u-s) = \frac{3}{2}[T_{\pi K}(s,t,u) + T_{\pi K}(u,t,s)] \qquad (38)$$

from which the amplitude entering the Lüscher formula follows by applying the forward scattering kinematics, $t=0$, viz.

$$\mathcal{F}_K(\tilde{\nu}) = T_{\pi K}^{I=0}(0, -4M_K\nu) \ . \qquad (39)$$

4.2.2 Kaon decay constant

We need the matrix element of an axial current between a kaon and two pions, which occurs in the evaluation of the K_{l4} decay. Ref. [24] defines the three scalar form factors F, G, R through

$$(A_K)_\mu = \frac{1}{\sqrt{2}} \langle \pi^+(p_1)\pi^-(p_2)|A_\mu^{4-i5}(0)|K^+(p)\rangle \qquad (40)$$

$$(A_K)_\mu = \frac{-i}{\sqrt{2}M_K}\Big[(p_1+p_2)_\mu F + (p_1-p_2)_\mu G + Q_\mu R\Big] \qquad (41)$$

with $F = F(s,t,u), G = G(s,t,u), R = R(s,t,u)$ and the kinematic variables

$$s = (p_1+p_2)^2 \ , \quad t = (p_1-p)^2 \ , \quad u = (p_2-p)^2 \ , \quad Q = p - p_1 - p_2 \ . \qquad (42)$$

The combination with the pions in an s-channel isospin zero state is

$$(A_K^{I=0})_\mu = \frac{i}{\sqrt{2}M_K}\Big[(p_1+p_2)_\mu F^+ + (p_1-p_2)_\mu G^- + Q_\mu R^+\Big] \qquad (43)$$

where

$$X^\pm = \frac{1}{2}[X(s,t,u) \pm X(s,u,t)]$$

---

[4] As noted in the literature, there are two typos in (3.16) of [25]: The prefactor of $(M_{\pi K}^r(u) - M_{K\eta}^r(u))$ should read $(M_K^2 - M_\pi^2)^2$, and the term multiplying $\frac{3}{8}J_{K\eta}^r(u)$ is $(u - \frac{2}{3}(M_\pi^2 + M_K^2))^2$ [the latter is correct in the preprint].

with $X = F, G, R$. Subtracting the pole at $Q^2 = (p - p_1 - p_2)^2 = M_K^2$ as defined in (25) and evaluating the amplitude $\bar{A}_K^{I=0}(s, t-u)$ in the forward scattering configuration $s = 0$ yields

$$\bar{A}_K^{I=0}(0, -4M_K\nu) = -\frac{3i}{\sqrt{2}M_K}\left[2\nu g^-(\nu) + M_K \bar{r}^+(\nu)\right] \tag{44}$$

with

$$g^-(\nu) = G^-(0, M_\pi^2 + M_K^2 - 2M_K\nu, M_\pi^2 + M_K^2 + 2M_K\nu) \tag{45}$$

and analogously for $\bar{r}^+$. Here, the bar on the form factor R^+ indicates again that it is defined after subtraction of the kaon pole (the form factor G^- remains unaffected by the subtraction). Finally, according to Eq.(24), the amplitude entering the Lüscher formula is

$$\mathcal{N}_K(\tilde{\nu}) = -i\bar{A}_K^{I=0}(0, -4M_K\nu) \ . \tag{46}$$

4.3 Eta

The decay constant of the η is of no phenomenological interest, like for the other neutral members of the pseudoscalar octet. We therefore refrain from discussing the finite volume effects for this quantity (as a side remark, we notice that also the analogue of the K_{l4} decay amplitude, the $\langle (2\pi)_{I=0}|A_\mu^8|\eta\rangle$ amplitude, is of no phenomenological interest and has never been calculated). We restrict ourselves to the finite volume effects on the η mass.

The $\pi\eta$-scattering amplitude $T_{\pi\eta}(s,t,u)$ has been calculated to NLO in Ref. [28],

$$\pi^0(p_1) + \eta(p_2) \to \pi^0(p_3) + \eta(p_4) \tag{47}$$

with kinematic variables (29), and relates to the isospin zero amplitude in the t-channel through

$$T_{\pi\eta}^{I=0}(t, u-s) = 3T_{\pi\eta}(s,t,u) \ . \tag{48}$$

The amplitude entering the Lüscher formula follows by applying the forward scattering kinematics, $t = 0$, viz.

$$\mathcal{F}_\eta(\tilde{\nu}) = T_{\pi\eta}^{I=0}(0, -4M_\eta\nu) \ . \tag{49}$$

5 Summary of the analytical results

In order to use the asymptotic formulae we have to feed them with the specific expressions for the scattering amplitude or the axial vector matrix element that are available in the literature. In this section we present such explicit formulae for $M_P(L)$ and $F_P(L)$ in two versions. We start with the complete expressions with some of the lengthier parts relegated to the appendix. The second step entails simplified versions of the unhandy parts, together with a discussion of how they relate to the complete version.

5.1 Full formulae

Evaluating (31) in the $SU(2)$ framework the fractional shift of the pion mass takes the form (26) with

$$
\begin{aligned}
I^{(2)}_{M_\pi} &= -B^0 \\
I^{(4)}_{M_\pi} &= B^0\left[-\frac{55}{18} + 4\bar{\ell}_1 + \frac{8}{3}\bar{\ell}_2 - \frac{5}{2}\bar{\ell}_3 - 2\bar{\ell}_4\right] \\
&\quad + B^2\left[\frac{112}{9} - \frac{8}{3}\bar{\ell}_1 - \frac{32}{3}\bar{\ell}_2\right] + S^{(4)}_{M_\pi} \\
I^{(6)}_{M_\pi} &= B^0\left[\frac{10049}{1296} - \frac{13}{72}N + \frac{20}{9}\bar{\ell}_1 - \frac{40}{27}\bar{\ell}_2 - \frac{3}{4}\bar{\ell}_3 - \frac{110}{9}\bar{\ell}_4 - \frac{5}{2}\bar{\ell}_3^2 - 5\bar{\ell}_4^2\right. \\
&\quad + \left(16\bar{\ell}_1 + \frac{32}{3}\bar{\ell}_2 - 11\bar{\ell}_3\right)\bar{\ell}_4 \\
&\quad + \ell_\pi\left(\frac{70}{9}\ell_\pi + 12\bar{\ell}_1 + \frac{32}{9}\bar{\ell}_2 - \bar{\ell}_3 + \bar{\ell}_4 + \frac{47}{18}\right) \\
&\quad \left. + 5\tilde{r}_1 + 4\tilde{r}_2 + 8\tilde{r}_3 + 8\tilde{r}_4 + 16\tilde{r}_5 + 16\tilde{r}_6\right] \\
&\quad + B^2\left[\frac{3476}{81} - \frac{77}{288}N + \frac{32}{9}\bar{\ell}_1 + \frac{464}{27}\bar{\ell}_2 + \frac{448}{9}\bar{\ell}_4 - \frac{32}{3}(\bar{\ell}_1 + 4\bar{\ell}_2)\bar{\ell}_4\right. \\
&\quad \left. + \ell_\pi\left(\frac{100}{9}\ell_\pi + \frac{8}{3}\bar{\ell}_1 + \frac{176}{9}\bar{\ell}_2 - \frac{248}{9}\right) - 8\tilde{r}_3 - 56\tilde{r}_4 - 48\tilde{r}_5 + 16\tilde{r}_6\right] \\
&\quad + S^{(6)}_{M_\pi} \quad (50)
\end{aligned}
$$

where we use the abbreviations (K_i denotes the modified Bessel function)

$$I_{M_P}^{2/4/6} \equiv I_{M_P}^{2/4/6}(\sqrt{n}\lambda_\pi), \qquad B^{0/2} \equiv B^{0/2}(\sqrt{n}\lambda_\pi) \qquad (51)$$

$$B^0(x) = 2K_1(x), \qquad B^2(x) = 2K_2(x)/x \qquad (52)$$

$$\ell_P = \ln(M_P^2/\mu^2), \qquad N = 16\pi^2 \qquad (53)$$

and the $\bar{\ell}_i$ which carry a mild logarithmic quark mass dependence [29]

$$\bar{\ell}_i = \bar{\ell}_i^{\text{phys}} + 2\log\left(\frac{M_\pi^{\text{phys}}}{M_\pi}\right) \qquad (54)$$

with mass independent $\bar{\ell}_i^{\text{phys}}$ as given in tab. 2. The terms $S_{M_\pi}^{(4)}, S_{M_\pi}^{(6)}$ are contributions from the loop functions at order p^4, p^6. They are explicitly given in app. A and their numerical importance is discussed below. Formula (50) has already appeared in [14].

The fractional shift in the pion decay constant takes the form (27) with

$$I_{F_\pi}^{(2)} = -2B^0$$

$$I_{F_\pi}^{(4)} = B^0\left[-\frac{7}{9} + 2\bar{\ell}_1 + \frac{4}{3}\bar{\ell}_2 - 3\bar{\ell}_4\right] + B^2\left[\frac{112}{9} - \frac{8}{3}\bar{\ell}_1 - \frac{32}{3}\bar{\ell}_2\right] + S_{F_\pi}^{(4)} \qquad (55)$$

and $S_{F_\pi}^{(4)}$ moved to the appendix. Formula (55) has already been given in [15].

With the abbreviation $x_{PQ} = M_P^2/M_Q^2$ the finite volume shift of the kaon mass is given by

$$I_{M_K}^{(2)} = 0$$

$$\begin{aligned}I_{M_K}^{(4)} &= 3x_{\pi K}^{1/2}\bigg\{B^0\bigg[\frac{x_{\pi K}}{9} + 8Nx_{\pi K}(4L_1^r + L_3^r - 4L_4^r - L_5^r + 4L_6^r + 2L_8^r) \\ &+ \ell_\pi\frac{x_{\pi K}^2}{4(1-x_{\pi K})} + \frac{\ell_K}{16}\bigg(-\frac{4}{1-x_{\pi K}} + \frac{1-10x_{\pi K}+x_{\pi K}^2}{6(x_{\eta K}-1)} + \frac{7+x_{\pi K}}{2}\bigg) \\ &+ \frac{\ell_\eta}{32}\bigg(\frac{2}{3} + (1-x_{\pi K})(x_{\eta K}-1) + \frac{53}{9}x_{\pi K} - \frac{x_{\pi K}^2}{3} - \frac{1-10x_{\pi K}+x_{\pi K}^2}{3(x_{\eta K}-1)}\bigg)\bigg] \\ &+ B^2\bigg[-8Nx_{\pi K}(4L_2^r + L_3^r) - \ell_\pi\frac{5x_{\pi K}^2}{2(1-x_{\pi K})} \\ &+ \ell_K\frac{x_{\pi K}}{2}\bigg(\frac{5}{1-x_{\pi K}} - \frac{1}{x_{\eta K}-1}\bigg) + \ell_\eta\frac{x_{\pi K}x_{\eta K}}{2(x_{\eta K}-1)}\bigg]\bigg\} + S_{M_K}^{(4)} \qquad (56)\end{aligned}$$

and the finite volume correction for F_K takes the form (27) with

$$I^{(2)}_{F_K} = -\frac{3}{4}B^0 \tag{57}$$

$$\begin{aligned}I^{(4)}_{F_K} &= B^0\bigg[\frac{3}{16}x_{\pi K}\bigg(\frac{\ell_K - x_{\pi K}\ell_\pi}{1 - x_{\pi K}} + \frac{\ell_\eta - x_{K\eta}\ell_K}{1 - x_{K\eta}} + 2\ell_\pi\Big(x_{\pi\eta} - \frac{9}{4}\Big)\bigg) \\&\quad + \frac{3}{32}(2\ell_K + 3x_{\eta K}\ell_\eta) + 12Nx_{\pi K}(4L^r_1 + L^r_3 - 2L^r_4) \\&\quad - 3NL^r_5(1 + x_{\pi K})\bigg] \\&\quad + B^2 x_{\pi K}\bigg[\frac{15}{2}\frac{\ell_K - x_{\pi K}\ell_\pi}{1 - x_{\pi K}} + \frac{3}{2}\frac{\ell_\eta - x_{K\eta}\ell_K}{1 - x_{K\eta}} - 24N(4L^r_2 + L^r_3)\bigg] \\&\quad + S^{(4)}_{F_K}\end{aligned} \tag{58}$$

and $S^{(4)}_{M_K}, S^{(4)}_{F_K}$ given in app. A and the convention (51) applied throughout, i.e. $I^{(4)}_{X_K} \equiv I^{(4)}_{X_K}(\sqrt{n}\lambda_\pi)$ and $S^{(4)}_{X_K} \equiv S^{(4)}_{X_K}(\sqrt{n}\lambda_\pi)$. Analogously, the eta mass in finite volume is given through

$$I^{(2)}_{M_\eta} = x^{3/2}_{\pi\eta}B^0$$

$$\begin{aligned}I^{(4)}_{M_\eta} &= x^{3/2}_{\pi\eta}\bigg\{B^0\bigg[-\frac{2 + x_{\pi\eta}}{3} + \ell_\pi x_{\pi\eta}\Big(\frac{2}{3(1 - x_{\pi\eta})} - \frac{13}{6}\Big) \\&\quad + \ell_K(2x_{K\eta} - x_{\pi\eta}) + \ell_\eta\Big(\frac{x_{\pi\eta}}{6} - \frac{2}{3(1 - x_{\pi\eta})}\Big) \\&\quad + 16N\Big(6(L^r_1 - L^r_4 + L^r_6 - L^r_7) + L^r_3 - L^r_5 + x_{\pi\eta}(6L^r_7 + 3L^r_8)\Big)\bigg] \\&\quad + B^2\Big[9(1 + \ell_K) - 32N(3L^r_2 + L^r_3)\Big]\bigg\} + S^{(4)}_{M_\eta} \,.\end{aligned} \tag{59}$$

Comparing (56) and (59) to (50) it is obvious that the $SU(3)$ breaking renders the expressions for $M_K(L), M_\eta(L)$ substantially more complicated than for $M_\pi(L)$. It is remarkable that to leading order the finite volume correction for the kaon mass vanishes [in the theory with virtual pions only, cf. eq. (99) in app. C], while for the eta mass there is a suppression factor $(M_\pi/M_\eta)^2$ [cf. eq. (100) in app. C]. As we shall see in the numerical analysis, these finite volume corrections are practically negligible.

5.2 Simplified formulae

Beyond tree-level, the chiral representation of the amplitudes $\mathcal{F}_P(\tilde{\nu}), \mathcal{N}_P(\tilde{\nu})$ that enter the resummed asymptotic formulae tend to become rather complicated. As a result, the expressions $S_{M_P}^{(4/6)}$ and $S_{F_P}^{(4)}$ are not particularly handy, see app. A. In the immediate vicinity of $\nu=0$, however, a polynomial approximation to the chiral amplitudes reproduces them rather well. The reason behind is that the nonanalytic structure closest to the origin is the cut starting at $\nu = \pm M_\pi$. Therefore, for imaginary ν close to the origin [i.e. for $y \in [-M_\pi, M_\pi]$ with y relating to ν via (14)] even a second order polynomial reproduces the amplitude rather accurately, while outside this region the quality of the representation does not matter, due to the suppression factor $\exp(-\sqrt{n(M_\pi^2 + y^2)}L)$. The advantage of such a polynomial representation is that all integrals can be performed analytically, and there is an analytic bound on the remainder.

For the pion mass and decay constant we find

$$\begin{aligned} S_{M_\pi}^{(4)} &= \frac{13}{3} g_0 B^0 - \frac{1}{3}\left(40 g_0 + 32 g_1 + 26 g_2\right) B^2 + O\!\left(B^4(\sqrt{n}\lambda_\pi)\right) \\ S_{F_\pi}^{(4)} &= \frac{1}{6}\left(8 g_0 - 13 g_1\right) B^0 - \frac{1}{3}\left(40 g_0 - 12 g_1 - 8 g_2 - 13 g_3\right) B^2 \\ &\quad + O\!\left(B^4(\sqrt{n}\lambda_\pi)\right) \end{aligned} \tag{60}$$

where the coefficients g_i are the Taylor coefficients of the function [5]

$$g(x) = \sigma \log \frac{\sigma-1}{\sigma+1} + 2 \tag{61}$$

with $\sigma = \sqrt{1-4/x}$ around the point $x=2$

$$g(2+\epsilon) = g_0 + g_1\epsilon + \frac{1}{2}g_2\epsilon^2 + \frac{1}{6}g_3\epsilon^3 + O(\epsilon^4) \tag{62}$$

with the explicit values

$$g_0 = 2 - \frac{\pi}{2}, \quad g_1 = \frac{\pi}{4} - \frac{1}{2}, \quad g_2 = \frac{1}{2} - \frac{\pi}{8}, \quad g_3 = \frac{3\pi}{16} - \frac{1}{2}. \tag{63}$$

For $S_{M_\pi}^{(6)}$ a similar short-hand version follows in the same manner. We refrain from showing them here, because these contributions turn out to be numerically so small that one could simply drop them. In $S_{M_K}^{(4)}, S_{F_K}^{(4)}$ and $S_{M_\eta}^{(4)}$ a polynomial

---

[5] The function $g(x)$ relates to the standard scalar one-loop function $\bar{J}$ through $g(x) = 16\pi^2 \bar{J}(xM_\pi^2)$.

i	$\bar{\ell}_i^{\text{phys}}$
1	-0.4 ± 0.6
2	4.3 ± 0.1
3	2.9 ± 2.4
4	4.4 ± 0.2

i	$\tilde{r}_i(M_\rho)$
1	$-1.5 \times (1 \pm 1)$
2	$3.2 \times (1 \pm 1)$
3	$-4.2 \times (1 \pm 1)$
4	$-2.5 \times (1 \pm 1)$
5	3.8 ± 1.0
6	1.0 ± 0.1

i	$L_i^{\text{r}}(M_\rho) \cdot 10^3$
1	0.38 ± 0.18
2	1.59 ± 0.15
3	-2.91 ± 0.32
4	0.00 ± 0.80
5	1.46 ± 0.10
6	0.00 ± 0.30
7	-0.49 ± 0.24
8	1.00 ± 0.21
9	6.90 ± 0.70

Tab. 2. $SU(2)$-framework: values at the physical pion mass of the low energy constants $\bar{\ell}_i$ from [29] together with the p^6 low energy constants $\tilde{r}_i(\mu = M_\rho)$. Note that the $\bar{\ell}_i$ used in the formulae for R_{M_π}, R_{F_π} differ from these values by a term logarithmic in $M_\pi/M_\pi^{\text{phys}}$, see (54). $SU(3)$-framework: $L_i^{\text{r}}(\mu = M_\rho)$ taken from the $O(p^4)$ fit in [30] (the uncertainties in L_6 and L_9 are our estimate).

expansion would not really simplify the representation, due to the different meson masses involved in the loop functions.

6 Numerical analysis

We are now in a position to evaluate the formulae for the relative finite volume corrections to M_P and F_P, as presented in the previous section. To fully specify the meaning of our formulae we need to give, as the last ingredient, the quark mass dependence of the infinite-volume quantities $M_\pi, M_K, M_\eta, F_\pi, F_K$. In line with the setup of our calculation we use 2-flavor ChPT for the pion mass and decay constant and 3-flavor ChPT for the kaon and eta counterparts. In either case the low energy parameters are determined from phenomenology and summarized in tab. 2. Regarding the $SU(2)$ low energy constants $\bar{\ell}_i$ and $\tilde{r}_i(\mu)$ we use the values obtained in Ref. [29, 22]; for the $SU(3)$ low energy constants $L_i^{\text{r}}(\mu)$ we refer to the $O(p^4)$ fit of Ref. [30]. In the former case, the full correlation matrix is given in [29] in the latter case it has been communicated privately [31], but in either case the final errors are almost the same, regardless whether the full correlation matrix is used or just the diagonal part.

6.1 M_π dependence of F_π

The quark mass dependence of M_π and F_π has been computed, up to 2 loops, in Refs. [32, 33, 29]. What we need in the present context is F_π as a function of M_π, i.e. the single relationship that one gets after eliminating the quark mass. This relation has been given in [14], where also the $SU(2)$ low energy constant $F = (86.2 \pm 0.5)$ MeV has been found. We stress that with M_π (or F_π) we mean simultaneously the pion mass (decay constant) in an infinite volume lattice simulation and in ChPT to the highest loop order available, i.e. to $O(p^6)$ in the $SU(2)$ framework. Finally, we mention that the chiral expansion parameter $\xi_\pi \equiv (M_\pi/4\pi F_\pi)^2$ remains small for pion masses up to 500 MeV, see the discussion in [14] for details.

6.2 M_π and m_s dependence of M_K, F_K and M_η

In the 3-flavor case we have two independent quark masses, the average down and up quark mass and the strange quark mass. We take the liberty to rewrite everything in terms of M_π and m_s, for reasons that will become obvious soon. Then the 1-loop quark mass formulae read

$$M_K^2 = \mathring{M}_K^2 + \frac{1}{F_\pi^2}\left[M_\pi^4\left\{-2k_1 + \frac{1}{4N}(-\ell_\pi + \frac{1}{3}\mathring{\ell}_\eta)\right\}\right.$$
$$\left. + m_s B_0(M_\pi^2 + m_s B_0)\left\{8(k_1 + 2k_2) + \frac{4}{9N}\mathring{\ell}_\eta\right\}\right] \quad (64)$$

$$M_\eta^2 = \mathring{M}_\eta^2 + \frac{1}{F_\pi^2}\left[M_\pi^4\left\{\frac{16}{9}(-k_1 + 2k_3) + \frac{1}{3N}(-2\ell_\pi + \mathring{\ell}_K)\right\}\right.$$
$$+ M_\pi^2 m_s B_0\left\{\frac{64}{9}(k_1 + 3k_2 - 2k_3) + \frac{4}{3N}(\mathring{\ell}_K - \frac{2}{9}\mathring{\ell}_\eta)\right\}$$
$$\left. + (m_s B_0)^2\left\{\frac{128}{9}(k_1 + \frac{3}{2}k_2 + k_3) + \frac{4}{3N}(\mathring{\ell}_K - \frac{8}{9}\mathring{\ell}_\eta)\right\}\right] \quad (65)$$

$$F_K = F_\pi + \frac{1}{F_\pi}\left[4(M_K^2 - M_\pi^2)L_5^r\right.$$
$$\left. + \frac{1}{N}\left\{\frac{5}{8}M_\pi^2\ell_\pi - \frac{1}{4}M_K^2\ell_K - \frac{3}{8}M_\eta^2\ell_\eta\right\}\right] \quad (66)$$

with

$$k_1 = 2L_8^r - L_5^r, \quad k_2 = 2L_6^r - L_4^r, \quad k_3 = 3L_7^r + L_8^r$$

and with the abbreviations

$$\overset{\circ}{M}{}_K^2 = m_s B_0 + \frac{1}{2} M_\pi^2 \,, \qquad \overset{\circ}{M}{}_\eta^2 = \frac{1}{3}(M_\pi^2 + 4 m_s B_0) \tag{67}$$

together with N and ℓ_P as defined in (53) and accordingly $\overset{\circ}{\ell}_P = \ln(\overset{\circ}{M}{}_P^2/\mu^2)$. Note that $\overset{\circ}{M}{}_K^2$ and $\overset{\circ}{M}{}_\eta^2$ are of hybrid nature – the M_π^2 part refers to 1-loop ChPT, while the part linear in m_s is a tree-level contribution. This is unavoidable if we want to discuss the dependence of physical quantities on the pion mass, instead of on quark masses. In technical terms (64) is used to fix $m_s B_0$; we simply require that M_K takes the physical value for $M_\pi = M_\pi^{\text{phys}}$, the result is $B_0 m_s = 0.223\,\text{GeV}^2$. The analogous requirement for M_η implies that we must slightly readjust L_7^r to $-0.47 \cdot 10^{-3}$ [well compatible with the error in tab. 2], for all other low energy constants the central values in tab. 2 are used. Even in the 3-flavor case we choose to describe the quark mass dependence of M_π and F_π through $SU(2)$ ChPT, as discussed in the previous subsection. Note that for $m_s = m_s^{\text{phys}}$ this choice exactly reproduces what one would get in the $SU(3)$ framework, since the phenomenological $\bar{\ell}_i$ values know about the virtual strange quark loops. In actual lattice simulations m_s is typically close to the physical value, and we expect this to remain a valid approximation. The resulting M_π dependence of F_K, M_K, M_η is shown in fig. 1. One notices that $M_\eta \sim 640\,\text{MeV}$ for $M_\pi \sim 500\,\text{MeV}$, thus the expansion parameter ξ_P remains small in the entire mass range $M_\pi \leq 500\,\text{MeV}$, even for $P = \eta$. In our numerical analysis we will use ξ_P exactly as determined from fig. 1 and ignore the uncertainty of this computed expansion parameter, since in a lattice computation one may iteratively determine $M_P(L)$ and $F_\pi(L)$ and thus ξ_P. We do, however, consider the uncertainties in the expansion coefficients $I_{M_P}^{(2/4/6)}$ and $I_{F_P}^{(2/4)}$ in (26) and (27), respectively, with details specified below. We have checked that even including the contribution of ξ_P to the total error would barely change the errors in our main result, figs. 2-5. In summary, the quark mass dependence of a quantity $X = F_K, M_K, M_\eta$ is considered a function of M_π alone through

$$X = X(M_\pi, m_s B_0 = 0.223\,\text{GeV}^2) \tag{68}$$

and an appropriate choice of the $SU(3)$ low energy constants, as given in tab. 2.

6.3 Results

We plot our results for R_{M_π} in fig. 2, both for $L = 2, 3, 4\,\text{fm}$ as a function of M_π (top) and for $M_\pi = 100, 300, 500\,\text{MeV}$ as a function of L (bottom). The result of the original ("$n=1$") Lüscher formula (13) with LO/NLO/NNLO chiral input is shown with a thin dotted/dashed/full line, respectively. The resummed ("all n") formula

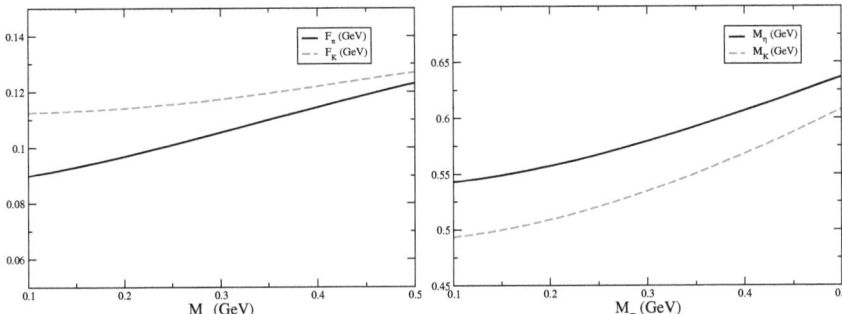

Fig. 1. M_π dependence (in infinite volume) of F_π, F_K, M_K, M_η. For the latter three quantities the strange quark mass has been fixed as to reproduce M_K^{phys} at $M_\pi = M_\pi^{\text{phys}}$.

(19) with LO/NLO/NNLO chiral input is given with a thick dotted/dashed/full line, respectively. With NLO or NNLO input the pertinent low energy constants lead to a non-negligible error band, except for R_{F_π} where the error with NLO input is of the order of the thickness of the line, with LO input it is zero. This is a consequence of our choice to disregard any uncertainty in the expansion parameter (here ξ_π), as discussed in the previous subsection. The area which corresponds – in the resummed scenario with NNLO input – to a situation with $M_\pi L \leq 2$ should be disregarded, since there is no reason to hope that the resummed Lüscher formula would still capture the numerically dominating terms in a complete "ChPT in finite volume" formula to the corresponding order in the p-counting.

Fig. 3 contains our data for $-R_{F_\pi}$. They are organized in the same manner as in the previous figure, though only LO and NLO input is used, since the pertinent matrix element is known only to 1-loop order, as discussed in Sect. 4.

Finally, figs. 4-5 contain the same information for F_K, M_K, M_η. In all these cases the asymptotic formulae with and without resummation may be compared, and the effect of going from LO to NLO input may be assessed.

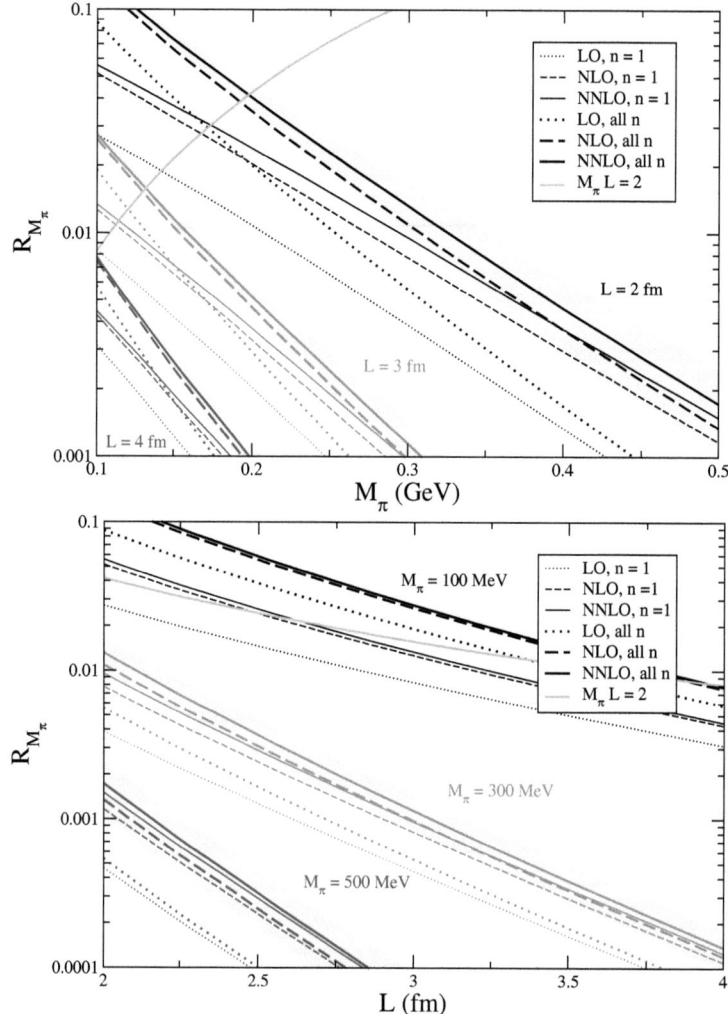

Fig. 2. R_{M_π} vs. M_π for $L = 2, 3, 4\,\text{fm}$ (top) and vs. L for $M_\pi = 100, 300, 500\,\text{MeV}$ (bottom). The result of the original ("$n=1$") Lüscher formula (13) with LO/NLO/NNLO chiral input is to be compared to the resummed ("all n") formula (19) which amounts to an approximate 1/2/3-loop ChPT calculation in finite volume. With NNLO input the low energy constants lead to a non-negligible error band; with NLO input the error is smaller (not shown), with LO input it is zero (see text). In the region above the $M_\pi L = 2$ line one is not safely in the p-regime and our results should not be trusted.

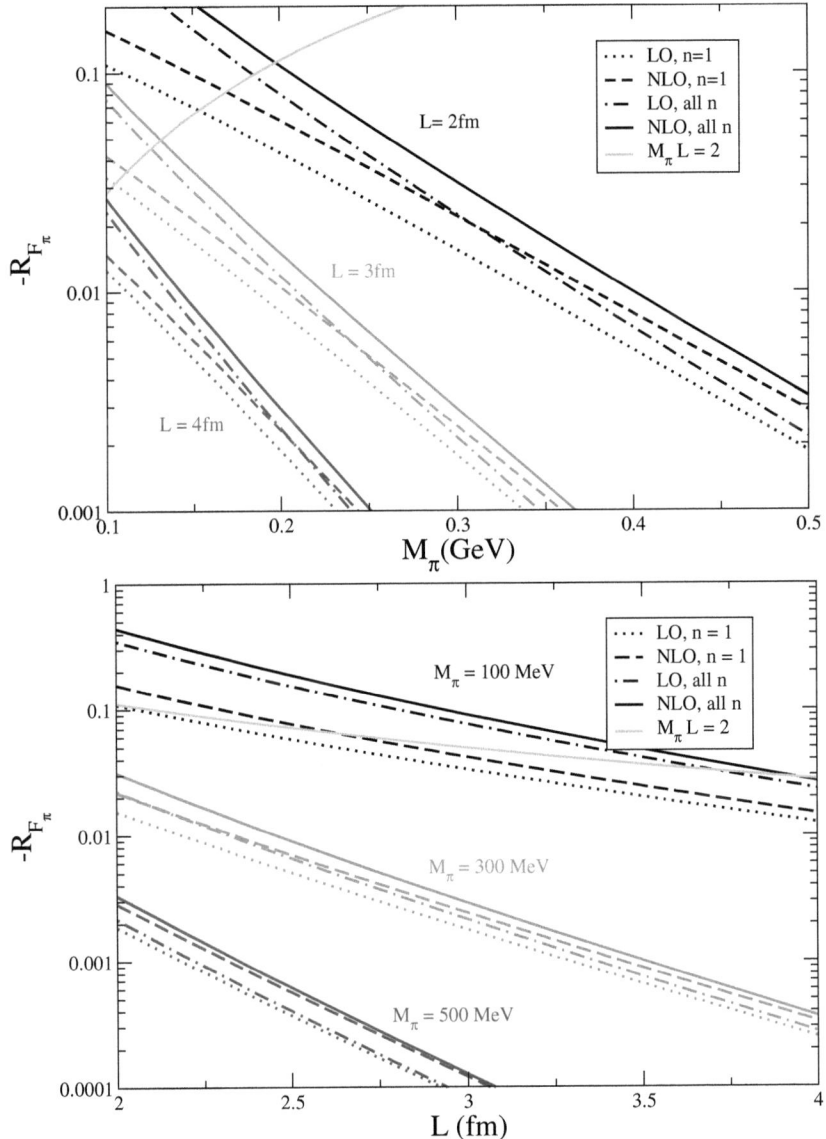

Fig. 3. $-R_{F_\pi}$ vs. M_π for $L = 2, 3, 4$ fm (left) and vs. L for $M_\pi = 100, 300, 500$ MeV (right).

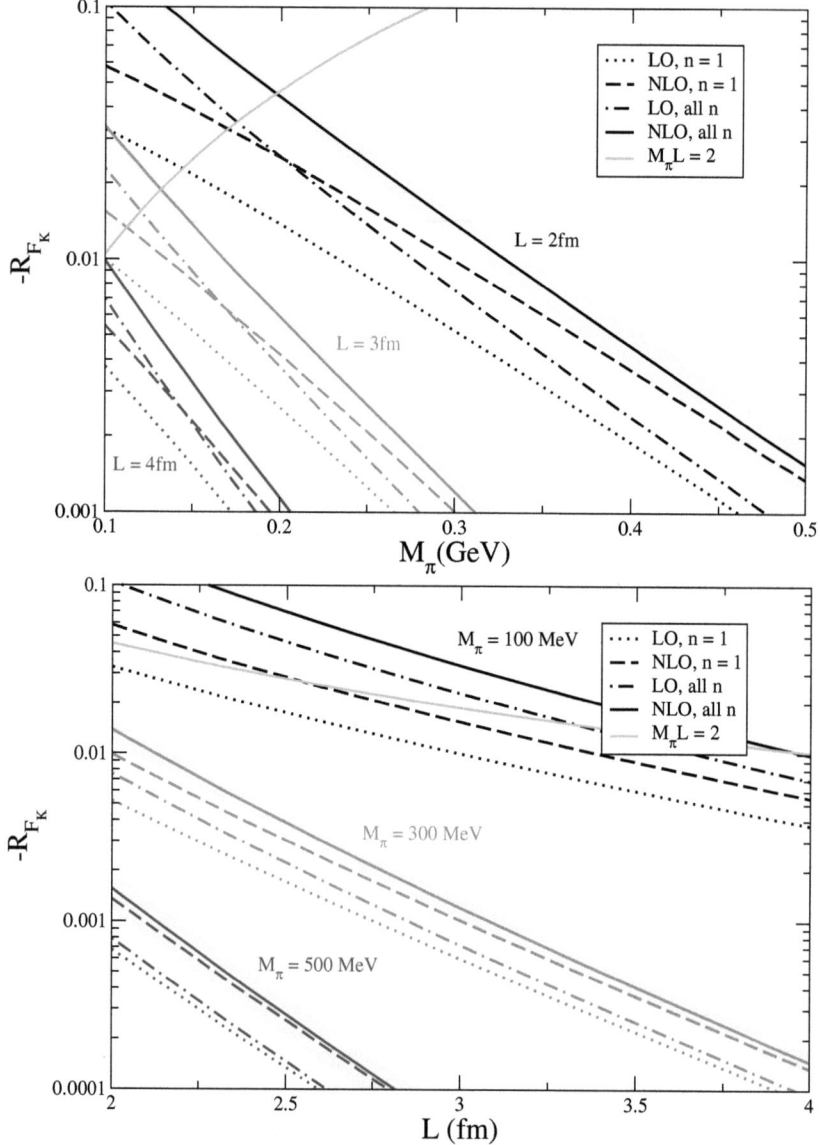

Fig. 4. $-R_{F_K}$ vs. M_π for $L = 2, 3, 4\,\text{fm}$ (left) and vs. L for $M_\pi = 100, 300, 500\,\text{MeV}$ (right).

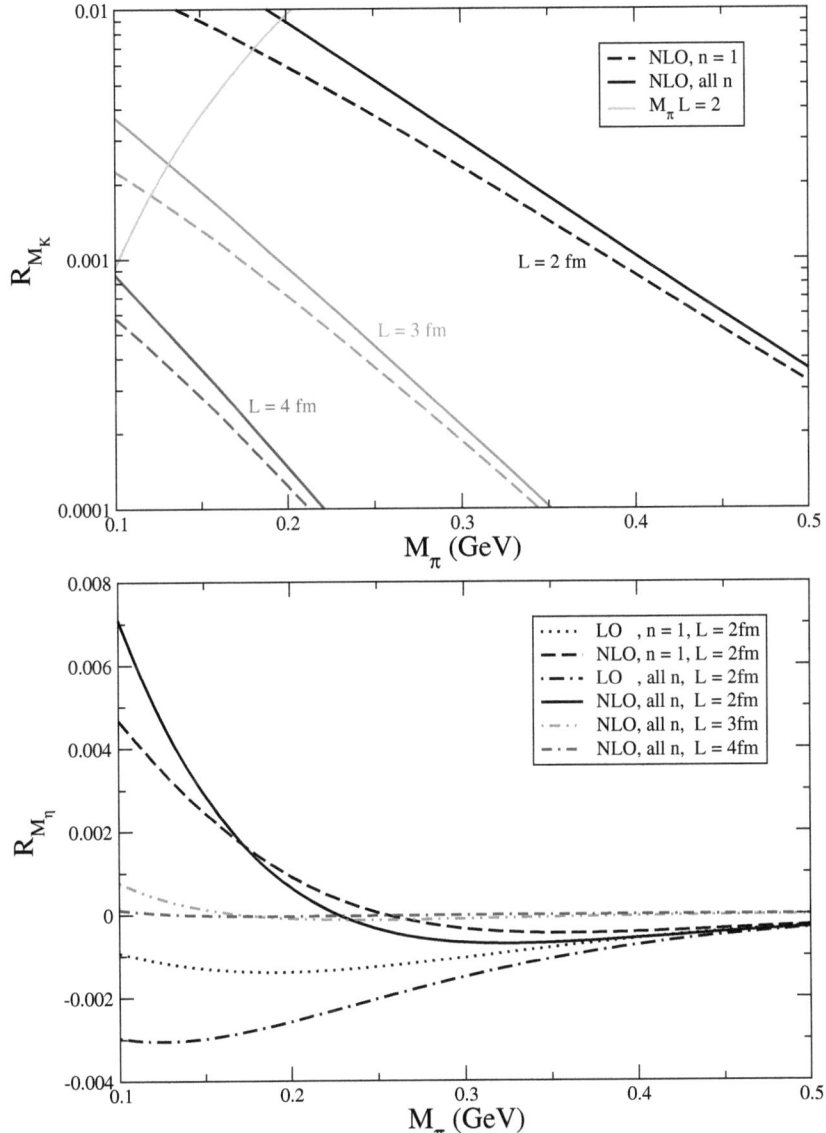

Fig. 5. R_{M_K} (left) and R_{M_η} (right) vs. M_π for $L=2,3,4$ fm.

R_{M_π}	1.6 fm	1.8 fm	2.0 fm	2.2 fm	2.4 fm	2.6 fm	2.8 fm	3.0 fm
140 MeV	.2099(127)	0.1292(82)	0.0830(55)	0.0552(38)	0.0377(27)	0.0264(19)	0.0188(14)	0.0136(10)
160 MeV	.1687(130)	0.1023(83)	0.0647(55)	0.0424(37)	0.0285(26)	0.0196(18)	0.0138(13)	0.0098(10)
180 MeV	.1366(130)	0.0817(82)	0.0509(53)	0.0328(35)	0.0217(24)	0.0147(17)	0.0102(12)	0.0071(09)
200 MeV	.1113(127)	0.0656(79)	0.0403(50)	0.0256(33)	0.0167(22)	0.0111(15)	0.0076(11)	0.0052(08)
220 MeV	.0912(123)	0.0529(75)	0.0320(47)	0.0200(31)	0.0129(20)	0.0084(14)	0.0056(09)	0.0038(06)
240 MeV	.0749(117)	0.0429(70)	0.0256(43)	0.0157(28)	0.0099(18)	0.0064(12)	0.0042(08)	0.0028(05)
260 MeV	.0618(110)	0.0349(65)	0.0205(39)	0.0124(25)	0.0077(16)	0.0049(10)	0.0032(07)	0.0021(05)
280 MeV	.0511(102)	0.0284(59)	0.0164(35)	0.0098(22)	0.0060(14)	0.0038(09)	0.0024(06)	0.0015(04)
300 MeV	0.0423(93)	0.0232(53)	0.0132(31)	0.0078(19)	0.0047(12)	0.0029(07)	0.0018(05)	0.0011(03)
320 MeV	0.0352(85)	0.0190(48)	0.0107(28)	0.0062(16)	0.0037(10)	0.0022(06)	0.0014(04)	0.0009(03)
340 MeV	0.0293(77)	0.0156(42)	0.0086(24)	0.0049(14)	0.0029(08)	0.0017(05)	0.0010(03)	0.0006(02)
360 MeV	0.0245(69)	0.0129(37)	0.0070(21)	0.0039(12)	0.0023(07)	0.0013(04)	0.0008(03)	0.0005(02)
380 MeV	0.0205(62)	0.0106(33)	0.0057(18)	0.0031(10)	0.0018(06)	0.0010(03)	0.0006(02)	0.0004(01)
400 MeV	0.0172(55)	0.0088(29)	0.0046(15)	0.0025(09)	0.0014(05)	0.0008(03)	0.0005(02)	0.0003(01)
420 MeV	0.0145(49)	0.0073(25)	0.0038(13)	0.0020(07)	0.0011(04)	0.0006(02)	0.0003(01)	0.0002(01)
440 MeV	0.0123(43)	0.0061(22)	0.0031(11)	0.0016(06)	0.0009(03)	0.0005(02)	0.0003(01)	0.0001(01)
460 MeV	0.0104(38)	0.0051(19)	0.0025(10)	0.0013(05)	0.0007(03)	0.0004(01)	0.0002(01)	0.0001(01)
480 MeV	0.0088(33)	0.0042(16)	0.0021(08)	0.0011(04)	0.0006(02)	0.0003(01)	0.0002(01)	0.0001(00)
500 MeV	0.0076(29)	0.0036(14)	0.0017(07)	0.0009(03)	0.0004(02)	0.0002(01)	0.0001(00)	0.0001(00)

Tab. 3. R_{M_π} via the resummed Lüscher formula (22) with NNLO chiral input for $\mathcal{F}_\pi(\tilde{\nu})$, representing an approximate 3-loop result. The error includes the uncertainty of the $\bar{\ell}_i$ and the $O(p^6)$ low energy constants, but no systematics. Entries with $M_\pi L < 2$ are unlikely to really capture the physical finite size effect, and the first two columns are somewhat on the short side with respect to the condition (4).

$-R_{F_\pi}$	1.6 fm	1.8 fm	2.0 fm	2.2 fm	2.4 fm	2.6 fm	2.8 fm	3.0 fm
140 MeV	0.5844(43)	0.3683(24)	0.2414(14)	0.1633(09)	0.1134(06)	0.0804(04)	0.0581(03)	0.0426(02)
160 MeV	0.4551(36)	0.2828(20)	0.1827(12)	0.1218(07)	0.0833(05)	0.0581(03)	0.0413(02)	0.0298(02)
180 MeV	0.3580(30)	0.2194(17)	0.1397(10)	0.0917(06)	0.0618(04)	0.0424(03)	0.0297(02)	0.0211(01)
200 MeV	0.2840(25)	0.1715(14)	0.1076(08)	0.0696(05)	0.0461(03)	0.0312(02)	0.0215(02)	0.0150(01)
220 MeV	0.2267(22)	0.1350(12)	0.0834(07)	0.0531(05)	0.0347(03)	0.0231(02)	0.0156(01)	0.0107(01)
240 MeV	0.1820(19)	0.1068(11)	0.0650(06)	0.0408(04)	0.0262(03)	0.0171(02)	0.0114(01)	0.0077(01)
260 MeV	0.1467(16)	0.0848(09)	0.0509(05)	0.0314(03)	0.0198(02)	0.0128(01)	0.0084(01)	0.0056(01)
280 MeV	0.1188(14)	0.0677(08)	0.0399(05)	0.0243(03)	0.0151(02)	0.0095(01)	0.0061(01)	0.0040(01)
300 MeV	0.0965(13)	0.0541(07)	0.0315(04)	0.0188(03)	0.0115(02)	0.0072(01)	0.0045(01)	0.0029(00)
320 MeV	0.0786(11)	0.0434(06)	0.0249(04)	0.0146(02)	0.0088(01)	0.0054(01)	0.0033(01)	0.0021(00)
340 MeV	0.0642(10)	0.0350(05)	0.0197(03)	0.0114(02)	0.0067(01)	0.0040(01)	0.0025(00)	0.0015(00)
360 MeV	0.0527(09)	0.0282(05)	0.0156(03)	0.0089(02)	0.0052(01)	0.0031(01)	0.0018(00)	0.0011(00)
380 MeV	0.0433(08)	0.0228(04)	0.0124(02)	0.0070(01)	0.0040(01)	0.0023(00)	0.0014(00)	0.0008(00)
400 MeV	0.0356(07)	0.0185(04)	0.0099(02)	0.0055(01)	0.0031(01)	0.0017(00)	0.0010(00)	0.0006(00)
420 MeV	0.0294(06)	0.0150(03)	0.0079(02)	0.0043(01)	0.0024(01)	0.0013(00)	0.0008(00)	0.0004(00)
440 MeV	0.0244(05)	0.0122(03)	0.0063(01)	0.0034(01)	0.0018(00)	0.0010(00)	0.0006(00)	0.0003(00)
460 MeV	0.0202(05)	0.0100(02)	0.0051(01)	0.0027(01)	0.0014(00)	0.0008(00)	0.0004(00)	0.0002(00)
480 MeV	0.0169(04)	0.0082(02)	0.0041(01)	0.0021(01)	0.0011(00)	0.0006(00)	0.0003(00)	0.0002(00)
500 MeV	0.0141(04)	0.0067(02)	0.0033(01)	0.0017(00)	0.0009(00)	0.0004(00)	0.0002(00)	0.0001(00)

Tab. 4. $-R_{F_\pi}$ via the resummed Lüscher formula (23) with NLO chiral input for $\mathcal{N}_\pi(\tilde{\nu})$, representing an approximate 2-loop result. The error includes the uncertainty of the $\bar{\ell}_i$, but no systematics. Entries with $M_\pi L < 2$ are unlikely to really capture the physical finite size effect, and the first two columns are somewhat on the short side with respect to the condition (4).

The main message to be extracted from figs. 2-5 is that the relative finite volume shift vanishes indeed in proportion to $e^{-M_\pi L}$; in the logarithmic representation one has an almost linear fall-off pattern, and higher orders mainly affect the prefactor. The second point concerns the relative importance of higher orders in the chiral counting versus higher exponentials. For small pion masses (say 100 MeV) resumming (i.e. higher exponentials) prove vital, while for large pion masses (say 500 MeV) higher loop orders prove more relevant (though the effect is small in that regime). As has been discussed in Ref. [14] a large shift in R_{M_P}, when moving from LO to NLO input, does not necessarily signal a bad chiral convergence behavior, since the cut in the underlying $\mathcal{F}_P$ amplitude starts only at the NLO level. Unfortunately, the effect due to an upgrade to NNLO input can only be checked in the case of R_{M_π}, since only there the pertinent amplitude is known. Nonetheless, we believe that the regime in the (M_π, L) plane that leads to a nice convergence behavior in R_{M_π} is indicative of the regime where the resummed formulae for $R_{F_\pi}, R_{M_K}, R_{F_K}, R_{M_\eta}$ with NLO input yield a trustworthy result.

For R_{M_π} and R_{F_π} the numerical results to the highest loop order available (equivalent to an approximate 3-loop and 2-loop calculation in ChPT) have been collected in tabs. 3 and 4, respectively. From the general discussion it is clear that the logarithms of the numbers in these tables may be interpolated with a low-order polynomial.

7 Two types of applications

We finish with a discussion of two prototype applications of our formulae. The first one is a "forward-type" application, in which our formulae are used to control a systematic error in a lattice calculation. The second one concerns a "backward-type" application, where one tries to determine QCD low energy constants from explicitly measuring finite-volume effects.

7.1 Finite volume effects and Marciano's determination of V_{us}

An example of how an analytic finite volume calculation may help to control a systematic error is the following. Marciano pointed out that, modulo radiative corrections, the ratio $\frac{V_{us}}{V_{ud}}\frac{F_K}{F_\pi}$ is fixed by the ratio of branching ratios for $K_{\ell 2}$ and $\pi_{\ell 2}$ decays. Taking into account radiative corrections, he obtained the following relation [34]

$$\frac{|V_{us}|^2}{|V_{ud}|^2}\frac{F_K^2}{F_\pi^2} = 0.07602(23)(27) \tag{69}$$

where the errors represent the experimental and radiative correction uncertainties. He then suggested to combine the value for V_{ud} obtained from superallowed nuclear beta decays with a value for F_K/F_π from lattice simulations. We stress that the necessary accuracy to make an impact on the determination of V_{us} is at the level of 1% or better, and indeed, both the determination of V_{ud} as well as the ratio of branching ratios are known to well below 1%. This means that any improvement in the lattice calculation of F_K/F_π will be immediately reflected in the value of V_{us}. In particular, being able to control systematic effects to well below 1% is of crucial importance.

With our results for R_{F_π} and R_{F_K} it is straightforward to calculate the finite-volume shift of the latter ratio

$$\frac{F_K(L)}{F_\pi(L)} = \frac{F_K}{F_\pi}\left\{1 + R_{F_K} - R_{F_\pi} + \mathcal{O}(R_F^2)\right\} \tag{70}$$

and thus to compare the magnitude of this effect to the typical size of the statistical error. A plot of the finite volume effect for the ratio of decay constants as a function of the pion mass and for a few volume sizes is provided in fig. 6. In his analysis Marciano uses the MILC Collaboration result $F_K/F_\pi = 1.201(8)(15)$ [35, 36]. Among the various data sets they have, those with the smallest $M_\pi L$ and thus most likely to be affected by sizeable finite volume corrections, have $M_\pi(L) \simeq 311\,\text{MeV}$, $L \simeq 2.4\,\text{fm}$ and $M_\pi(L) \simeq 262\,\text{MeV}$, $L \simeq 2.89\,\text{fm}$. Using (70) we find that with the parameters of the first set $F_K(L)/F_\pi(L)$ in the continuum deviates from the infinite volume result by $0.0099 - 0.0038 = 0.61\%$, while the corresponding estimate for the second set reads $0.0067 - 0.0025 = 0.42\%$. A systematic effect of half a percent needs to be taken into account in a high precision study, and from fig. 6 one sees that other (M_π, L) pairs to reach that level would be $(410\,\text{MeV}, 2\,\text{fm})$, $(230\,\text{MeV}, 3\,\text{fm})$ and $(150\,\text{MeV}, 4\,\text{fm})$.

We stress that the numerical example just discussed is for illustrative purposes only, because we do not know whether our formulae can be applied to the MILC Collaboration data. Lattice QCD with staggered fermions and $N_f = 2$ or $N_f = 2+1$ violates flavour (or taste) symmetry and low energy unitarity, properties our analysis relies on. Under the assumption that these effects disappear in the continuum limit [6], the finite volume shift of an observable like F_π/F_K can be calculated in

---

[6] As of now, there is no proof, but rather a lively debate on this issue in the literature [37].

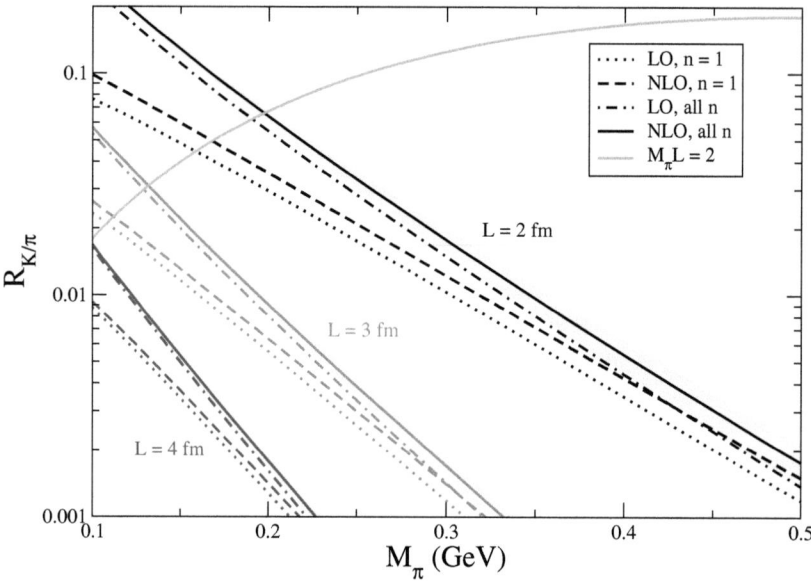

Fig. 6. The relative finite volume effect $(F_K(L)F_\pi)/(F_\pi(L)F_K) - 1$ vs. M_π for $L = 2, 3, 4$ fm. Above the $M_\pi L = 2$ line one is not safely in the p-regime and our results should not be trusted.

staggered chiral perturbation theory (this is what the MILC Collaboration does [38]), but one cannot enjoy the benefits of the Lüscher formula. We stress that no such conceptual issues arise with dynamical Wilson-type fermions. In such a case our continuum formulae can be directly applied to the data at finite lattice spacing with cut-off effects bringing only mild (i.e. numerically irrelevant) modifications as discussed in app. B.

7.2 Low energy constants from finite volume effects

In the present section we discuss whether one can use these finite volume effects to obtain information on the low energy constants from lattice calculations. At first sight this seems unpractical, because these effects are quite small and decay exponentially with $M_\pi L$. This means that, roughly speaking, if one wishes to obtain information on (a combination of) low energy constants to a certain accuracy, one has to calculate the corresponding particle mass or decay constant to an accuracy which is about two orders of magnitude higher, and this is a challenge. The asymptotic formulae provide a connection between finite volume effects on two-point functions and (infinite volume) four-point functions, and thus give us access to low energy constants that appear as local contributions to four-point functions. The question is then what alternative ways one has to measure these constants on the lattice. It is well known that the $P\pi$ scattering amplitudes which govern the finite volume corrections to the mass of the P particle can be obtained more directly by evaluating the finite volume dependence of the energy of the state of two particles P and π enclosed in a box [39]. This method is more direct because the effect is suppressed only by powers of the volume rather than exponentially. Still, such a calculation is very difficult, also because the typical volume needed is quite large.

There is another important difference between the two methods. As we have seen in Sect. 5.2 the analytic representation of the finite volume effects is enormously simplified if one Taylor expands the amplitudes in the Lüscher-type formulae. Here, two terms in the Taylor expansion are enough for a very accurate representation. The extraction of the low energy constants from these effects can be viewed as a two-step process: one first determines the values of the first two Taylor coefficients of the amplitudes at $\nu = 0$, and then from these the low energy constants. The chiral representation enters only in this second step. Analogously, if one uses the method with two particles in a box, one first determines the scattering lengths, and then extracts from these the relevant low energy constants. As discussed in [29], the scattering lengths have a badly converging chiral expansion, such that only at very small quark masses one would be able to reliably extract the low energy constants. This happens because one is evaluating the amplitude on top of the threshold singularity. At $\nu = 0$, below threshold and away from any other singularity, the amplitude displays a better convergence. For all these reasons we believe it is worthwhile to explore this alternative route.

The quantities which are worth considering for our scope are M_π, F_π and F_K. We write the relative finite volume shifts in the form

$$R_X = R_X^0 + c_X \left(\beta^0 L_X^0 + \beta^2 L_X^2 \right) \tag{71}$$

where R_X^0 represents contributions independent of the low energy constants, and the coefficients c_X are defined as

$$c_{M_\pi} = -\xi_\pi^2, \quad c_{F_\pi} = \xi_\pi^2, \quad c_{F_K} = 12N \frac{F_\pi}{F_K} \xi_\pi \xi_K . \tag{72}$$

The functions $\beta^{0,2}$ are series of Bessel functions and depend on λ_π only

$$\beta^{0,2} = \sum_{n=1}^{\infty} \frac{m(n)}{\sqrt{n}\lambda_\pi} B^{0,2}(\sqrt{n}\lambda_\pi) . \tag{73}$$

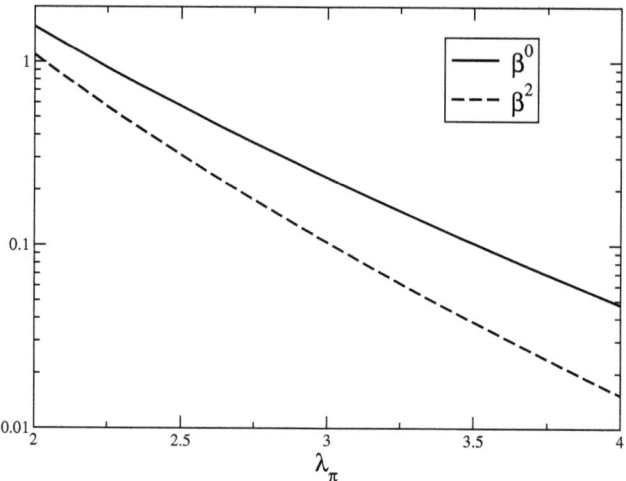

Fig. 7. The functions $\beta^{0,2}$.

They are plotted in fig. 7. The $L_X^{0,2}$ are the combinations of low energy constants that appear in each of the quantities to next-to-leading order. In particular we read off from the formulae (22,23) and (50,55,58)

$$\begin{aligned}
L^0_{M_\pi} &= 2(\bar{\ell}_1 + \frac{2}{3}\bar{\ell}_2) - \frac{5}{4}\bar{\ell}_3 - \bar{\ell}_4 \\
L^2_{M_\pi} &= -\frac{4}{3}(\bar{\ell}_1 + 4\bar{\ell}_2) \\
L^0_{F_\pi} &= 2(\bar{\ell}_1 + \frac{2}{3}\bar{\ell}_2) - 3\bar{\ell}_4 \\
L^2_{F_\pi} &= -\frac{8}{3}(\bar{\ell}_1 + 4\bar{\ell}_2) \\
L^0_{F_K} &= x_{\pi K}(4L^r_1 + L^r_3 - 2L^r_4) - \frac{1}{4}(1 + x_{\pi K})L^r_5 \\
L^2_{F_K} &= -2x_{\pi K}(4L^r_2 + L^r_3) \, .
\end{aligned} \qquad (74)$$

It is interesting that both M_π and F_π are sensitive to the same combination of $\bar{\ell}_1$ and $\bar{\ell}_2$ (note that $\bar{\ell}_3$ and $\bar{\ell}_4$ can be pinned down directly from the quark mass dependence of M_π and F_π, respectively). This means that one can determine the low energy constants which fix the $\pi\pi$ scattering amplitude also from the finite volume effects for F_π where the $\pi\pi$ scattering amplitude does not appear. One should remark that the sensitivity to the combination $(\bar{\ell}_1 + \frac{2}{3}\bar{\ell}_2)$ is the same, whereas F_π is a factor two more sensitive to the combination $(\bar{\ell}_1 + 4\bar{\ell}_2)$.

To discuss the numerics we fix $M_\pi = 300$ MeV and $L = 2$ fm. In this setting $\beta^0 = 0.22$, $\beta^2 = 0.10$, and $c_{M_\pi} = -0.0026$, $c_{F_\pi} = 0.0026$. A change of one unit in the combination $(\bar{\ell}_1 + \frac{2}{3}\bar{\ell}_2)$ then generates a shift of about 0.12% in both $M_\pi(L)$ and $F_\pi(L)$. According to tab. 2 this linear combination is known from phenomenology to be

$$(\bar{\ell}_1 + \frac{2}{3}\bar{\ell}_2)|_{M_\pi = 300\text{MeV}} = -0.1 \pm 0.6 \ . \tag{75}$$

On the other hand a change of one unit in $(\bar{\ell}_1 + 4\bar{\ell}_2)$ modifies $F_\pi(L)$ by 0.07% and $M_\pi(L)$ by half that much. This linear combination, however, is known to be larger from phenomenology

$$(\bar{\ell}_1 + 4\bar{\ell}_2)|_{M_\pi = 300\text{MeV}} = 9.1 \pm 0.5 \ . \tag{76}$$

This numerical example indicates that in order to get a reasonable account on these combinations of low energy constants, one would have to control the pion mass and decay constant for $M_\pi = 300$ MeV and $L = 2$ fm to less than 1 permille, which is a real challenge.

An alternative way, as already mentioned, would be to calculate the $\pi\pi$ S-wave $I = 2$ scattering lengths via Lüscher's method [39]. There are some results with dynamical fermions for this quantity [40], but not yet precise enough and for low enough pion masses that would allow an extraction of the relevant low energy constant. We mention in passing that both a_0^2 and a_0^0 (the latter is even more difficult to calculate on the lattice, and shows a very badly converging chiral expansion [29]) are sensitive to the combination $\bar{\ell}_1 + 2\bar{\ell}_2$, and therefore provide complementary information to the one that would be obtained from the finite volume effects we have discussed here.

The numerics for the F_K case is similar. Again, in order to get a sensitivity comparable to the one which characterizes the phenomenological determination (notice that the L_i^r constants are usually given in units of 10^{-3}) one would have to calculate F_K to the permille accuracy.

8 Conclusions

In this paper we have carefully analyzed the finite volume effects on masses and decay constants of the lightest pseudoscalar mesons. The theoretical framework in which we carry out this analysis has been set up long ago by Gasser and Leutwyler [4] and by Lüscher [12]. The former two discussed how ChPT can be adapted to the case of a finite box and then be used to calculate finite volume effects. The latter author derived a formula valid for large volumes which expresses the finite volume effects in terms of an integral over a physical scattering amplitude. This formula does not rely on ChPT, but for QCD is best used in combination with ChPT: this is the tool to provide a representation of the necessary scattering amplitude as a function of quark masses. More recently a formula à la Lüscher for decay constants has been derived by two of us [15]. In the present paper we have proposed a resummation of the asymptotic formulae as the best tool to study finite volume effects for masses and decay constants. The resummation is a simple, "kinematical" extension of the asymptotic formulae, which however does improve the algebraic accuracy of the formula. Used in combination with the chiral representation for the scattering amplitude in the integral it yields an accurate determination of finite volume effects.

Our numerical results show that decay constants and masses of the pseudoscalars are in general little affected by the finite spatial size of the box (as soon as the box is large enough that ChPT can be applied, $L \geq 2\,\text{fm}$). For the smallest acceptable values of $M_\pi L$ (in the p-regime, in which we are working, this quantity has to be larger than one) they are typically of the order of a few percent for M_π, F_π and F_K. We have seen that in the p-regime M_K and M_η are practically insensitive to the box size. Independently of the exact size of these corrections we could always check the convergence of the chiral expansion and conclude that these finite volume effects are under good theoretical control. This means that if one's goal is to calculate masses or decay constants one can use the results of this paper to choose the volume in order to minimize the calculational costs. For example, by using a box of 2 fm size (which our results show to be sufficiently large) and explicitly correcting for the finite volume effects one can save computational costs with respect to a 2.5 fm size box which (for pion masses of about 300 MeV or larger) gives finite volume effects below 1%. The gain in CPU time that comes from such a reduction of the volume by almost a factor 2 can be used more fruitfully by pushing towards smaller lattice spacings and/or lighter quark masses.

Since our results are theory predictions, an explicit check by lattice calculations would of course be very welcome, but it would require a high precision. Once this level of accuracy will be reached, our formulae will become particularly useful. First, in order to correct for these effects in all applications which require a high precision, like the evaluation of the F_K/F_π ratio which, as suggested by Marciano [34], leads to a determination of V_{us} [38]. Second, if one wants to use these finite size effects as a means to determine low energy constants on the lattice. As we have discussed, despite the fact that these effects are exponentially suppressed and numerically quite small in all practical situations, they do offer the advantage of involving in two-point functions low energy constants which, otherwise, appear only in four-point functions. The latter are quantities which are very difficult to determine directly on the lattice, and it may therefore turn out to be easier to see them indirectly, through a small correction to an "easy" quantity like the pion mass or decay constant.

Acknowledgments

We are indebted to Peter Hasenfratz, Heiri Leutwyler and Rainer Sommer for useful discussions and/or comments on the manuscript. This work has been supported by the Schweizerischer Nationalfonds and partly by the EU "Euridice" program under code HPRN-CT2002-00311.

A The integrals $S_{M_P}^{(n)}$ and $S_{F_P}^{(n)}$

In this appendix we give explicitly the contributions from loop functions to the finite volume effects. These have been introduced and defined in Sect. 5. The expression for the pion related integrals have already been given in [14, 15]. We give them here for convenience:

$$S_{M_\pi}^{(4)} = \frac{13}{3}R_0^0 - \frac{16}{3}R_0^1 - \frac{40}{3}R_0^2$$

$$S_{M_\pi}^{(6)} = R_0^0\left(\frac{817}{27} + \frac{80}{9}\bar{\ell}_2 - 5\bar{\ell}_3 + \frac{52}{3}\bar{\ell}_4\right) - \frac{2}{3}R_0^1\left(\frac{313}{9} + \frac{40}{3}\bar{\ell}_2 + 32\bar{\ell}_4\right)$$

$$
\begin{aligned}
&+ R_0^2\left(\frac{292}{27} - 8\bar{\ell}_1 - \frac{128}{9}\bar{\ell}_2 - \frac{160}{3}\bar{\ell}_4\right) + \frac{4}{3}R_0^3\left(-\frac{47}{9} + 4\bar{\ell}_1 - 4\bar{\ell}_2\right) \\
&+ \frac{1}{9}R_1^0\left(1 - \frac{\pi^2}{2}\right) + \frac{1}{9}R_1^1\left(128 - \frac{\pi^2}{8}\right) - \frac{1}{3}R_1^2\left(\frac{100}{3} + \frac{\pi^2}{8}\right) \\
&+ \frac{1}{6}R_2^0\left(7 - \frac{\pi^2}{3}\right) + \frac{1}{9}R_2^1\left(16 + \frac{7\pi^2}{8}\right) + \frac{\pi^2}{24}R_2^2 \\
&- \frac{46}{9}R_3^0 - \frac{32}{9}R_3^1 - \frac{32}{3}R_3^2 + \frac{40}{3}\left(R_4^0 + R_4^1\right)
\end{aligned}
$$

$$S_{F_\pi}^{(4)} = \frac{4}{3}\left(R_0^0 - R_0^1 - 10R_0^2\right) - \frac{13}{6}R_0^{0\prime} + \frac{8}{3}R_0^{1\prime} + \frac{20}{3}R_0^{2\prime} \tag{77}$$

where the integrals R_i^k are defined as

$$R_0^{k(\prime)} \equiv R_0^{k(\prime)}(\sqrt{n}\lambda_\pi) = \begin{cases} \text{Re} \\ \text{Im} \end{cases} \int_{-\infty}^{\infty} d\tilde{y}\, \tilde{y}^k\, e^{-\sqrt{n(1+\tilde{y}^2)}\lambda_\pi}\, g^{(\prime)}(2+2i\tilde{y})$$

$$\text{for } \begin{cases} k \text{ even} \\ k \text{ odd} \end{cases} \tag{78}$$

$$R_i^k \equiv R_i^k(\sqrt{n}\lambda_\pi) = \begin{cases} \text{Re} \\ \text{Im} \end{cases} \int_{-\infty}^{\infty} d\tilde{y}\, \tilde{y}^k\, e^{-\sqrt{n(1+\tilde{y}^2)}\lambda_\pi}\, N^2\, K_i^{\pi\pi}(2+2i\tilde{y})$$

$$\text{for } \begin{cases} k \text{ even} \\ k \text{ odd} \end{cases} \tag{79}$$

with $g^{(\prime)}$ defined in (61) and

$$
\begin{aligned}
K_1^{\pi\pi}(x) &= \frac{1}{N^2\sigma^2}\Big[g(x) - 2\Big]^2 \\
K_2^{\pi\pi}(x) &= \frac{1}{N^2}\Big[g(x)^2 - 4g(x)\Big] \\
K_3^{\pi\pi}(x) &= \frac{1}{2N^2\sigma^4 x}\Big[2g(x)^3 - 12g(x)^2 + 24g(x) + 2\pi^2\sigma^2 g(x) - 16 - \pi^2\sigma^2 x\Big] \\
K_4^{\pi\pi}(x) &= \frac{1}{\sigma^2 x}\Big[K_0^{\pi\pi}(x) + \frac{1}{2}K_1^{\pi\pi}(x) + \frac{1}{3}K_3^{\pi\pi}(x) + \frac{(\pi^2-6)x}{12N^2}\Big]
\end{aligned}
\tag{80}
$$

with $\sigma = \sqrt{1 - 4/x}$. As seen in (78), in $S_{F_\pi}^{(4)}$ integrals over the derivative of the function g appear. This is a consequence of the subtraction procedure (34). In the practical implementation one writes $s_1 = 2M_\pi^2 - 2\nu M_\pi + (Q^2 - M_\pi^2)/2 - s_3/2$ and $s_2 = 2M_\pi^2 + 2\nu M_\pi + (Q^2 - M_\pi^2)/2 - s_3/2$ to trade s_1, s_2 for $Q^2 - M_\pi^2, \nu$ and expands $A_\pi^{I=0}$ consistently in these new variables. For instance in F_0 one substitutes $\bar{J}(s_1) \to \bar{J}(2M_\pi^2 - 2\nu M_\pi) + \bar{J}'(2M_\pi^2 - 2\nu M_\pi)[Q^2 - M_\pi^2 - s_3]/2$.

The loop integrals for kaon and eta finite volume effects read:

$$
\begin{aligned}
S_{M_K}^{(4)} &= 3\Bigg\{\frac{3}{32}(1+x_{\pi K})^2 S_{K\pi}^{0,1} - \frac{5}{8}(1+x_{\pi K})S_{K\pi}^{1,1} - \frac{19}{8}S_{K\pi}^{2,1} \\
&\quad - \frac{3}{16}(1-x_{\pi K}^2)S_{K\pi}^{0,3} + \frac{13}{8}(1-x_{\pi K})S_{K\pi}^{1,3} - \frac{3}{2}\left(x_{\pi K}S_{K\pi}^{0,5} + S_{K\pi}^{2,5}\right) \\
&\quad + \frac{1}{96}(1+x_{\pi K})^2 S_{\eta K}^{0,1} - \frac{1}{8}(1+x_{\pi K})S_{\eta K}^{1,1} - \frac{3}{8}S_{\eta K}^{2,1} \\
&\quad + \frac{1}{16}(1+x_{\pi K})(3x_{\eta K} + 2x_{\pi K} - 5)S_{\eta K}^{0,3} \\
&\quad + \frac{3}{8}(1 - 2x_{\pi K} + x_{\eta K})S_{\eta K}^{1,3} - \frac{3}{2}\left(x_{\pi K}S_{\eta K}^{0,5} + S_{\eta K}^{2,5}\right)\Bigg\} \\
S_{F_K}^{(4)} &= \frac{M_K}{M_\pi}\Bigg[-\frac{15}{16}(1+x_{\pi K})S_{K\pi}^{1,1} - \frac{57}{8}S_{K\pi}^{2,1} - \frac{9}{64}(1+x_{\pi K})^2 S_{K\pi}^{0,2} \\
&\quad + \frac{15}{16}(1+x_{\pi K})S_{K\pi}^{1,2} + \frac{57}{16}S_{K\pi}^{2,2} + \frac{9}{32}(1-5x_{\pi K})S_{K\pi}^{0,3} \\
&\quad + \left(6 - \frac{15}{8}x_{\pi K}\right)S_{K\pi}^{1,3} + \frac{15}{4}S_{K\pi}^{2,3} + \frac{9}{32}(1-x_{\pi K}^2)S_{K\pi}^{0,4} \\
&\quad - \frac{39}{16}(1-x_{\pi K})S_{K\pi}^{1,4} - \frac{9}{4}\left(x_{\pi K}S_{K\pi}^{0,5} + 2S_{K\pi}^{2,5} - x_{\pi K}S_{K\pi}^{0,6} - S_{K\pi}^{2,6}\right) \\
&\quad - \frac{3}{16}(1+x_{\pi K})S_{\eta K}^{1,1} - \frac{9}{8}S_{\eta K}^{2,1} - \frac{1}{64}(1+x_{\pi K})^2 S_{\eta K}^{0,2} \\
&\quad + \frac{3}{16}(1+x_{\pi K})S_{\eta K}^{1,2} + \frac{9}{16}S_{\eta K}^{2,2} \\
&\quad + \left(\frac{27}{32}(x_{\eta K} - 1) - \frac{3}{16}(1+x_{\pi K})\right)S_{\eta K}^{0,3} \\
&\quad + \frac{3}{8}(4 - 2x_{\pi K} + 3x_{\eta K})S_{\eta K}^{1,3} + \frac{9}{4}S_{\eta K}^{2,3} \\
&\quad + \frac{3}{32}(5 - 3x_{\eta K})(1 - x_{\pi K}^2)S_{\eta K}^{0,4} \\
&\quad - \frac{9}{16}(1 - 2x_{\pi K} + x_{\eta K})S_{\eta K}^{1,4} \\
&\quad - \frac{9}{4}\left(x_{\pi K}S_{\eta K}^{0,5} + 2S_{\eta K}^{2,5} - x_{\pi K}S_{\eta K}^{0,6} - S_{\eta K}^{2,6}\right)\Bigg]
\end{aligned}
$$

$$S_{M_\eta}^{(4)} = x_{\pi\eta}^2 T_{KK}^{0,1} - 6x_{\pi\eta} T_{KK}^{1,1} - 9T_{KK}^{2,1} + \frac{2}{3} x_{\pi\eta}^2 T_{\eta\pi}^{0,1} \tag{81}$$

where we have introduced the following abbreviations

$$x_{PQ} = \frac{M_P^2}{M_Q^2}, \qquad \ell_P = \ln\left(\frac{M_P^2}{\mu^2}\right). \tag{82}$$

The integrals B^{2k} are proportional to modified Bessel functions

$$\begin{aligned} B^{2k} \equiv B^{2k}(\sqrt{n}\lambda_\pi) &= \int_{-\infty}^{\infty} d\tilde{y}\, \tilde{y}^{2k}\, e^{-\sqrt{n(1+\tilde{y}^2)}\,\lambda_\pi} \\ &= \frac{\Gamma(k+1/2)}{\Gamma(3/2)} \left(\frac{2}{\sqrt{n}\lambda_\pi}\right)^k K_{k+1}(\sqrt{n}\lambda_\pi) \end{aligned} \tag{83}$$

and the quantities $S_{PQ}^{k,I}$ and $T_{PQ}^{k,I}$ are integrals over functions g_{PQ}^I which occur at one-loop order in the chiral expansion. They are all analytical along the integration line. The expressions $S_{PQ}^{k,I}$ and $T_{PQ}^{k,I}$ are defined as

$$S_{PQ}^{k,I} = \begin{cases} \text{Re} \\ \text{Im} \end{cases} N x_{\pi K}^{(k+1)/2} \int_{-\infty}^{\infty} d\tilde{y}\, \tilde{y}^k\, e^{-\sqrt{n(1+\tilde{y}^2)}\,\lambda_\pi} g_{PQ}^{(I)}(M_K^2 + M_\pi^2 + 2\mathrm{i} M_K M_\pi \tilde{y})$$

$$\text{for } \begin{cases} k \text{ even} \\ k \text{ odd} \end{cases}$$

$$T_{PQ}^{k,I} = \begin{cases} \text{Re} \\ \text{Im} \end{cases} N x_{\pi\eta}^{(k+1)/2} \int_{-\infty}^{\infty} d\tilde{y}\, \tilde{y}^k\, e^{-\sqrt{n(1+\tilde{y}^2)}\,\lambda_\pi} g_{PQ}^{(I)}(M_\eta^2 + M_\pi^2 + 2\mathrm{i} M_\eta M_\pi \tilde{y})$$

$$\text{for } \begin{cases} k \text{ even} \\ k \text{ odd} \end{cases}$$

with

$$\begin{aligned} g_{PQ}^{(1)}(x) &= \bar{J}_{PQ}(x), & g_{PQ}^{(2)}(x) &= M_K^2 \bar{J}'_{PQ}(x) \\ g_{PQ}^{(3)}(x) &= K_{PQ}(x), & g_{PQ}^{(4)}(x) &= M_K^2 K'_{PQ}(x) \\ g_{PQ}^{(5)}(x) &= \bar{M}_{PQ}(x), & g_{PQ}^{(6)}(x) &= M_K^2 \bar{M}'_{PQ}(x). \end{aligned}$$

For completeness, we give the explicit expressions for the g_{PQ}^I [24]. All functions can be expressed in terms of the subtracted scalar integral $\bar{J}(t) = J(t) - J(0)$ evaluated in four dimensions

$$J(t) = -\mathrm{i} \int \frac{d^d p}{(2\pi)^d} \frac{1}{((p+k)^2 - M^2)(p^2 - m^2)} \tag{84}$$

with $t = k^2$. The functions used in the text are then

$$\bar{J}(t) = -\frac{1}{N}\int_0^1 dx \,\ln\frac{M^2 - tx(1-x) - \Delta x}{M^2 - \Delta x}$$

$$= \frac{1}{2N}\left\{2 + \frac{\Delta}{t}\ln\frac{m^2}{M^2} - \frac{\Sigma}{\Delta}\ln\frac{m^2}{M^2} - \frac{\sqrt{\rho}}{t}\ln\frac{(t+\sqrt{\rho})^2 - \Delta^2}{(t-\sqrt{\rho})^2 - \Delta^2}\right\}$$

$$\bar{J}'(t) = -\frac{2}{N}\frac{M^2 m^2}{(t-\Sigma)^2 - \rho}\frac{1}{t^2}\left[2t + \Delta\ln\frac{m^2}{M^2} + \frac{t\Sigma - \Delta^2}{\sqrt{\rho}}\ln\frac{(t+\sqrt{\rho})^2 - \Delta^2}{(t-\sqrt{\rho})^2 - \Delta^2}\right]$$

$$\bar{K}(t) = \frac{\Delta}{2t}\bar{J}(t)$$

$$\bar{K}'(t) = -\frac{\Delta}{2t}\left(\frac{\bar{J}(t)}{t} - \bar{J}'(t)\right)$$

$$\bar{M}(t) = \frac{1}{12t}\{t - 2\Sigma\}\bar{J}(t) + \frac{\Delta^2}{3t^2}\bar{J}(t) + \frac{1}{18N} - \frac{1}{6Nt}\left\{\Sigma + 2\frac{M^2 m^2}{\Delta}\ln\frac{m^2}{M^2}\right\}$$

$$\bar{M}'(t) = \frac{1}{6t^2}\left[\frac{\Sigma t - 4\Delta^2}{t}\bar{J}(t) + \frac{1}{2}(t^2 - 2t\Sigma + 4\Delta^2)\bar{J}'(t)\right.$$

$$\left. + \frac{1}{N}\left(\Sigma + \frac{2M^2 m^2}{\Delta}\ln\frac{m^2}{M^2}\right)\right] \quad (85)$$

where

$$\Delta = M^2 - m^2, \quad \Sigma = M^2 + m^2$$

$$\rho = \rho(t, M^2, m^2) = (t+\Delta)^2 - 4tM^2 . \quad (86)$$

In the text these are used with subscripts

$$\bar{J}_{PQ}(t) = \bar{J}(t) \quad \text{with} \quad M = M_P, m = M_Q \quad (87)$$

and similarly for the other symbols. We add a remark concerning the analyticity properties of the loop functions. The asymptotic formula requires them to be evaluated for complex arguments. There is one case, where the representation of eq.(85) does not yet provide an unambiguous analytic continuation, namely for $\bar{J}_{PQ}(M_P^2 + M_Q^2 + 2iM_P M_Q \tilde{y})$, because $\rho = -4M_P^2 M_Q^2(1 + \tilde{y}^2)$. The correct analytical continuation is given by

$$\sqrt{\rho} = 2iM_P M_Q \omega \quad (88)$$

with $\omega = \sqrt{1+\tilde{y}^2}$, implying for the logarithm in eq.(85) (for $t = M_P^2 + M_Q^2 + 2iM_P M_Q \tilde{y}$),

$$\ln \frac{(t+\sqrt{\rho})^2 - \Delta^2}{(t-\sqrt{\rho})^2 - \Delta^2} = \ln \frac{\omega + \tilde{y}}{\omega - \tilde{y}} \begin{cases} +i\pi & \text{for } \tilde{y} < 0 \\ -i\pi & \text{for } \tilde{y} > 0 \end{cases}. \qquad (89)$$

All loop functions are now well defined along the integration line in the asymptotic formulae.

B Cut-off effects

In this appendix we wish to discuss whether it is sufficient to calculate finite volume effects in continuum ChPT or whether cut-off effects should be taken care of when correcting [7] actual lattice data for the effect of the finite spatial box length L.

Naive reasoning suggests that – because finite volume effects are due to the pion cloud around a particle and thus to pure IR physics – such shifts will be rather insensitive to the UV properties of the theory. This is what one expects to hold as long as the cut-off is large compared to the scale of chiral symmetry breaking, $\Lambda_{\text{XSB}} \simeq 1\,\text{GeV}$. With a lattice regularization all momenta are cut off at π/a, and with a standard lattice spacing $a \simeq 0.1\,\text{fm}$ the resulting scale $\sim 6\,\text{GeV}$ is indeed much bigger than Λ_{XSB}.

This intuitive argument can be refined in two ways. The first option is to invoke an extension of ChPT designed to take care of the effects of the finite lattice spacing a. Of course, the details of this theory need to be tailored to the action used, but generically the new Lagrangian follows from the old one by replacing the low energy constants, e.g. $\ell_3 \to \ell_3 + \text{const}\, w_3$ – see Ref. [41] for a recent review. In consequence, the $O(p^4)$ formula (8) takes the form

$$M_\pi(a,L) = M\left\{1 - \frac{1}{4}x\left[\tilde{\ell}_3 + \text{const}\,\tilde{w}_3 - \log(x)\right] + \frac{1}{2N_f} x \tilde{g}_1(\lambda_\pi) + O(x^2)\right\} \qquad (90)$$

where $x = M^2/(4\pi F)^2$, $M = \sqrt{2Bm}$ and $\tilde{\ell}_3 = \log(\Lambda_3^2/(4\pi F)^2)$. In other words the very effect of such an extension concerns the particle mass in infinite volume, the

---

[7] Here we assume that the finite volume correction is applied before the continuum and chiral extrapolations, thus interchanging steps (i) and (ii) of Sect. 1. In practice such a change is helpful, since otherwise the volumes would have to be matched to perform a well-defined continuum extrapolation and this would require a priori knowledge of the lattice spacing that will come out of the simulation.

fractional finite-size effect gets modified by a-effects only at $O(p^4)$ in the chiral counting, viz.

$$M_\pi(a,L) = M_\pi(a)\left(1 + \frac{1}{2N_f}x\tilde{g}_1(\lambda_\pi) + O(x^2)\right). \tag{91}$$

The second option is to compare the generic one loop finite volume shift in the continuum to a version in which the pion propagator is discretized in the simplest[8] possible way

$$g_1(M,L,0) = \int \frac{dp_0}{2\pi}\left\{\frac{1}{L^3}\sum_{\frac{2\pi}{L}Z^3}\frac{1}{\mathbf{p}^2+p_0^2+M^2} - \int\frac{d^3p}{(2\pi)^3}\frac{1}{\mathbf{p}^2+p_0^2+M^2}\right\} \tag{92}$$

$$g_1(M,L,a) = \int \frac{dp_0}{2\pi}\left\{\frac{1}{L^3}\sum_{\text{finite}}\frac{1}{\hat{\mathbf{p}}^2+p_0^2+M^2} - \int_{-\pi/a}^{+\pi/a}\frac{d^3p}{(2\pi)^3}\frac{1}{\hat{\mathbf{p}}^2+p_0^2+M^2}\right\} \tag{93}$$

[the definition $\hat{p}^2 = \frac{4}{a^2}\sum_i \sin^2(\frac{a}{2}p_i)$ has been used and the finite sum runs over $p_i = \frac{2\pi}{L}n_i$ with $n_i \in \{0,...,N-1\}$ and $N=L/a$ an integer] and verify that the difference is small compared to the shift itself, i.e. $|g_1(M,L,0)-g_1(M,L,a)| \ll g_1(M,L,0)$ for standard values of M,L,a. With

$$g_1(M,L,0) = \int_0^\infty dt\,\frac{1}{16\pi^2 t^2}\sum_{n\geq 1}m(n)\,e^{-nL^2/(4t)-M^2 t}$$

$$= \frac{1}{4\pi^2}\sum_{n\geq 1}m(n)\frac{M K_1(\sqrt{n}ML)}{\sqrt{n}L} \tag{94}$$

$$g_1(M,L,a) = \int_0^\infty dt\left(\left[\frac{1}{L}\sum_{n=0}^{N-1}\exp\left(-\frac{4t}{a^2}\sin^2\left(\frac{\pi n}{N}\right)\right)\right]^3 - \left[\frac{I_0(2t/a^2)}{a\,e^{2t/a^2}}\right]^3\right)\frac{e^{-tM^2}}{\sqrt{4\pi t}} \tag{95}$$

[cf. tab. 1 for $m(n)$] and $M=300$ MeV, $L=2$ fm, $a=0.1$ fm one finds

$$(4\pi/M)^2 g_1(M,L,0) = 0.4374 \tag{96}$$

and

$$(4\pi/M)^2 g_1(M,L,a) = 0.4400, \tag{97}$$

---

[8] By considering (93,95) we do not indicate that this would yield a better estimate of the finite size effects than the continuum form (92,94). The actual discretization of quarks and gluons will not lead to a simple pion propagator, but we are interested in the absolute difference $|g_1(M,L,0) - g_1(M,L,a)|$, since it is expected to correctly indicate the order of magnitude of discretization effects in the finite volume shift of actual data. We stress that our discretization is similar in spirit, but not identical to the one used in Ref. [42].

thus a difference of less than a percent in a quantity designed to correct actual data by – at most – a few percent. It is hence sufficient to calculate the finite volume effects in continuum ChPT. This conclusion was also reached in Ref. [8].

C Effects due to kaon and eta loops

Our formulae (22, 23) take only the effects due to virtual pion loops into account. In other words, they neglect the contribution to $M_P(L) - M_P$ and $F_P(L) - F_P$ coming from kaon and eta loops "around the world". In this appendix we want to discuss to which extent this is justified.

Consider a soon-to-be standard $N_f = 2+1$ simulation with $M_\pi = 300$ MeV and the strange quark fixed at its physical value and $M_K = 530$ MeV in consequence (see fig. 1). With a box size $L = 2$ fm a first crude estimate says that the kaon loop effects will be down, relative to the pion loops, by a factor $e^{-(M_K - M_\pi)L} = 0.1$, and a 10% correction on the fractional finite volume effect is not exactly small. However, this correction should be compared to the absolute size of the effect and the statistical error and from tabs. 3-4 and/or figs. 2-5 it follows that it is safe to neglect such a correction at the permille level. Finally, we mention that one expects NNLO contributions for R_{F_π}, R_{F_K} to be of the same order of magnitude as for R_{M_π} and this means that an additional pion loop could prove more important than a single kaon loop in finite volume.

We have verified that kaon and eta loops "around the world" prove numerically insignificant by comparing the pion-loop contributions to those of the kaons and etas in the full 1-loop expressions calculated in $SU(3)$ ChPT in finite volume. We find

$$R_{M_\pi} = \frac{1}{4}\xi_\pi \tilde{g}_1(\lambda_\pi) - \frac{1}{12}\xi_\eta \tilde{g}_1(\lambda_\eta) \tag{98}$$

$$R_{M_K} = \frac{1}{6}\xi_\eta \tilde{g}_1(\lambda_\eta) \tag{99}$$

$$R_{M_\eta} = \frac{1}{2}\xi_K \tilde{g}_1(\lambda_K) - \frac{1}{3}\xi_\eta \tilde{g}_1(\lambda_\eta)$$
$$+ \frac{M_\pi^2}{M_\eta^2}\left[-\frac{1}{4}\xi_\pi \tilde{g}_1(\lambda_\pi) + \frac{1}{6}\xi_K \tilde{g}_1(\lambda_K) + \frac{1}{12}\xi_\eta \tilde{g}_1(\lambda_\eta)\right] \tag{100}$$

where λ_P and $\tilde{g}_1$ have been defined in (11) and (12), respectively, and

$$R_{F_\pi} = -\xi_\pi \tilde{g}_1(\lambda_\pi) - \frac{1}{2}\xi_K \tilde{g}_1(\lambda_K) \qquad (101)$$

$$R_{F_K} = -\frac{3}{8}\xi_\pi \tilde{g}_1(\lambda_\pi) - \frac{3}{4}\xi_K \tilde{g}_1(\lambda_K) - \frac{3}{8}\xi_\eta \tilde{g}_1(\lambda_\eta) \qquad (102)$$

$$R_{F_\eta} = -\frac{3}{2}\xi_K \tilde{g}_1(\lambda_K) . \qquad (103)$$

These formulae deserve a few comments. First, a few elementary checks: Our (98), (99) and (102) agree with $\Delta M_\pi/M_\pi, \Delta M_K/M_K, \Delta F_K/F_K$ as given in [10, 17] and both the R_M and the R_F become degenerate in the $SU(3)$ limit ($m_u = m_d = m_s$), and furthermore, these degenerate expressions agree with the result by Gasser and Leutwyler [2], eqns. (8) and (9), specified to $N_f = 3$. Second, R_{M_K} and R_{F_η} have no $\tilde{g}(\lambda_\pi)$ (in other words only kaon- and eta-loops contribute at one-loop order) and in R_{M_η} the one-pion-loop contribution is suppressed by an extra factor M_π^2/M_η^2. Therefore, one expects $M_K(L) - M_K$, $M_\eta(L) - M_\eta$ and $F_\eta(L) - F_\eta$ to be small, and the numerical investigation in Sect. 6 specifies to which extent this is true. The numerical discussion also shows that, for a substantial range of pion masses, we have $M_\eta(L) < M_\eta$, in (apparent) contradiction to our statement in Sect. 2 that finite volume effects will lift the masses of the pseudo Goldstone bosons. However, this just indicates that the general rule may be overwhelmed by $SU(3)$ breaking effects. As our formulae show, the $SU(3)$ breaking effects are accidentally dominating (at one-loop level) only in the R_{M_P} and not in the R_{F_P}, i.e. $R_{F_P} < 0$ for $P = \pi, K, \eta$. Finally, we wish to elaborate on a point already raised in the work by Becirevic and Villadoro [10]. The main message of the one-loop formulae (98 - 103) is that finite volume effects and chiral logs are intimately related. In this particular case, where there is only a tadpole contribution, the finite volume shift follows from the quark mass dependence by the simple substitution $\log(M_P^2/\mu^2) \to \log(M_P^2/\mu^2) + \tilde{g}_1(\lambda_P)$. In practice, this means that one cannot extract "chiral logs" and the pertinent low energy constants without controlling the finite volume effects.

References

[1] G. Colangelo, Nucl. Phys. Proc. Suppl. 140 (2005) 120 [hep-lat/0409111].
[2] J. Gasser and H. Leutwyler, Phys. Lett. B 184, 83 (1987).
[3] J. Gasser and H. Leutwyler, Phys. Lett. B 188, 477 (1987).
[4] J. Gasser and H. Leutwyler, Nucl. Phys. B 307, 763 (1988).
[5] A. Ali Khan et al. [QCDSF-UKQCD Collaboration], Nucl. Phys. B 689, 175 (2004) [hep-lat/0312030].
[6] S.R. Beane, Phys. Rev. D 70, 034507 (2004) [hep-lat/0403015].
[7] S.R. Beane and M.J. Savage, Phys. Rev. D 70, 074029 (2004) [hep-ph/0404131].
[8] S.R. Sharpe, Phys. Rev. D 46, 3146 (1992) [hep-lat/9205020].
[9] J. Braun, B. Klein and H.J. Pirner, Phys. Rev. D 71 (2005) 014032 [hep-ph/0408116].
[10] D. Becirevic and G. Villadoro, Phys. Rev. D 69, 054010 (2004) [hep-lat/0311028].
[11] D. Arndt and C.J.D. Lin, Phys. Rev. D 70, 014503 (2004) [hep-lat/0403012].
[12] M. Lüscher, Commun. Math. Phys. 104, 177 (1986).
[13] Y. Koma and M. Koma, Nucl. Phys. B 713, 575 (2005) [hep-lat/0406034].
[14] G. Colangelo and S. Dürr, Eur. Phys. J. C 33, 543 (2004) [hep-lat/0311023].
[15] G. Colangelo and C. Haefeli, Phys. Lett. B 590, 258 (2004) [hep-lat/0403025].
[16] H. Neuberger, Phys. Rev. Lett. 60, 889 (1988). H. Neuberger, Nucl. Phys. B 300, 180 (1988). P. Hasenfratz and H. Leutwyler, Nucl. Phys. B 343, 241 (1990). F.C. Hansen, Nucl. Phys. B 345, 685 (1990). W. Bietenholz, Helv. Phys. Acta 66, 633 (1993) [hep-th/9402072].
[17] S. Descotes-Genon, Eur. Phys. J. C 40, 81 (2005) [hep-ph/0410233].
[18] W. Detmold and C.J.D. Lin, Phys. Rev. D 71, 054510 (2005) [hep-lat/0501007].
[19] G. Colangelo, S. Dürr and R. Sommer, Nucl. Phys. Proc. Suppl. 119, 254 (2003) [hep-lat/0209110].
[20] G. Colangelo and C. Haefeli, work in progress
[21] A. Schenk, Phys. Rev. D 47, 5138 (1993).
[22] J. Bijnens, G. Colangelo, G. Ecker, J. Gasser and M. E. Sainio, Phys. Lett. B 374, 210 (1996) [hep-ph/9511397]. J. Bijnens, G. Colangelo, G. Ecker,

J. Gasser and M.E. Sainio, Nucl. Phys. B 508, 263 (1997) [Erratum-ibid. B 517, 639 (1998)] [hep-ph/9707291].

[23] G. Colangelo, M. Finkemeier and R. Urech, Phys. Rev. D 54, 4403 (1996) [hep-ph/9604279].

[24] J. Bijnens, G. Colangelo and J. Gasser, Nucl. Phys. B 427, 427 (1994) [hep-ph/9403390].

[25] V. Bernard, N. Kaiser and U.G. Meissner, Nucl. Phys. B 357 129 (1991).

[26] J. Bijnens, P. Dhonte and P. Talavera, JHEP 0405, 036 (2004) [hep-ph/0404150].

[27] G. Amoros, J. Bijnens and P. Talavera, Nucl. Phys. B 585, 293 (2000) [Erratum-ibid. B 598, 665 (2001)] [hep-ph/0003258].

[28] V. Bernard, N. Kaiser and U. G. Meissner, Phys. Rev. D 44, 3698 (1991).

[29] G. Colangelo, J. Gasser and H. Leutwyler, Nucl. Phys. B 603, 125 (2001) [hep-ph/0103088].

[30] G. Amoros, J. Bijnens and P. Talavera, Nucl. Phys. B 602, 87 (2001) [hep-ph/0101127].

[31] J. Bijnens, private communication.

[32] J. Gasser and H. Leutwyler, Annals Phys. 158, 142 (1984).

[33] U. Bürgi, Nucl. Phys. B 479,392 (1996). [hep-ph/9602429].

[34] W.J. Marciano, Phys. Rev. Lett. 93, 231803 (2004) [hep-ph/0402299].

[35] C. Aubin et al. [MILC Collaboration], Nucl. Phys. Proc. Suppl. 129, 227 (2004) [hep-lat/0309088].

[36] C. Aubin et al., Phys. Rev. D 70, 094505 (2004) [hep-lat/0402030].

[37] B. Bunk, M. Della Morte, K. Jansen and F. Knechtli, Nucl. Phys. B 697, 343 (2004) [hep-lat/0403022]. A. Hart and E. Muller, Phys. Rev. D 70, 057502 (2004) [hep-lat/0406030]. C. Aubin and C. Bernard, Phys. Rev. D 68, 034014 (2003) [hep-lat/0304014]. S.R. Sharpe and R.S. Van de Water, [hep-lat/0409018]. S. Dürr and C. Hoelbling, Phys. Rev. D 69, 034503 (2004) [hep-lat/0311002]. E. Follana, A. Hart and C.T.H. Davies [HPQCD Collaboration], Phys. Rev. Lett. 93, 241601 (2004) [hep-lat/0406010]. S. Dürr, C. Hoelbling and U. Wenger, Phys. Rev. D 70, 094502 (2004) [hep-lat/0406027]. K.Y. Wong and R.M. Woloshyn, [hep-lat/0412001]. S. Dürr and C. Hoelbling, Phys. Rev. D 71, 054501 (2005) [hep-lat/0411022]. F. Maresca and M. Peardon, [hep-lat/0411029]. D.H. Adams, [hep-lat/0411030]. Y. Shamir, Phys. Rev. D 71, 034509 (2005) [hep-lat/0412014]. C. Bernard, Phys. Rev. D 71, 094020 (2005) [hep-lat/0412030].

[38] C. Aubin et al. [MILC Collaboration], Phys. Rev. D 70, 114501 (2004) [hep-lat/0407028].

[39] M. Lüscher, Commun. Math. Phys. 105, 153 (1986).

[40] T. Yamazaki et al. [CP-PACS Collaboration], Phys. Rev. D 70, 074513 (2004) [hep-lat/0402025].

[41] O. Bär, Nucl. Phys. Proc. Suppl. 140, 106 (2005) [hep-lat/0409123].

[42] B. Borasoy and R. Lewis, Phys. Rev. D 71, 014033 (2005) [hep-lat/0410042].

III

Finite volume effects for the pion mass at two-loops

Finite volume effects for the pion mass at two-loops[1]

Christoph Haefeli

Institut für Theoretische Physik, Universität Bern
Sidlerstr. 5, 3012 Bern, Switzerland

Abstract

We evaluate the pion mass in finite volume to two loops within Chiral Perturbation Theory. The results are compared with a recently proposed extension of the asymptotic formula of Lüscher. We find that contributions, which were neglected in the latter, are numerically very small at the two-loop level and argue that for $M_\pi L \gtrsim 2$, $L \geq 2\text{fm}$ the finite volume effects in the meson sector are analytically well under control.

1 Introduction

Numerical simulations with lattice QCD are bound to rather small lattice volumes when determining the hadron spectrum and other low energy parameters in QCD. In principle the computed observables show a volume dependence and a thorough understanding of these effects is important for a correct interpretation of numerical data. In the case of the pion mass, Lüscher has been able to establish a precise asymptotic formula which relates the size dependence of the pion mass with the $\pi\pi$ forward scattering amplitude [1]. The contributions in the low energy region are enhanced, where the $\pi\pi$ scattering amplitude is represented very accurately through its chiral representation [2, 3, 4, 5]. The combined use of the asymptotic formula and Chiral Perturbation Theory (ChPT) led to a quantitative investigation [6, 7] which confirmed the expectation that the Lüscher formula provides an efficient method to exploit finite volume effects. Even more, the pion mass is one of those observables, where a high precision measurement on the lattice should finally

---
[1] In collaboration with Gilberto Colangelo

be possible. If one is able to evaluate finite size effects of the pion mass on the lattice, the asymptotic formula might be used for a determination of the scattering amplitude. For this to be true, the asymptotic formula must be extended. As we will discuss, the derivation of the asymptotic formula relies on various approximations and takes into account only the leading exponential contributions of the order $\exp(-M_\pi L)$ and systematically drops contributions of the order $\exp(-\sqrt{3}M_\pi L/2)$. The question then arises whether the missed terms might turn out to be numerically relevant. The same question also concerns the asymptotic formulae of decay constants [8]. A first step to address this question has been performed by one of us in Ref.[9]. There, it is proposed to extend the Lüscher formula in a straightforward way: one resummes the most important part of higher order contributions. The so obtained resummed formula is then exact on the one-loop level. However, since the resummation does not take into account all higher order terms, it still remains to be justified, in what sense this can be considered as an improvement. The present article investigates this conjecture with a full two-loop calculation of the pion mass in finite volume ChPT [10, 11, 12]. The results confirm our expectation that contributions which were omitted in the asymptotic approach are very small for $M_\pi L \gtrsim 2$ at the two-loop level. This implies that the resummed formula combined with ChPT indeed provides a convenient laboratory for a determination of low energy constants from lattice QCD.

The two-loop calculation appears to be interesting also in its own right. To date, a number of finite volume calculations have been performed at one-loop order [13, 14, 15, 16, 17, 18], but as far as we know two-loop calculations have only been performed for the quark condensate [19] and for low-energy observables in the related context of finite temperature field theory [20, 21]. As finite volume effects occur first at the one-loop level, only a two-loop calculation – or alternatively if existing, an asymptotic formula à la Lüscher – leads to a better understanding of the convergence behaviour of the perturbative expansion. The results of the present two-loop calculation were already presented in [22]. Here, we give further details and report on the derivations.

We wish to briefly mention related work. Most notably, the asymptotic formula may also be applied to the nucleon mass [23], see Koma and Koma [24] and Colangelo and Fuhrer [25] for recent work. Braun, Pirner and Klein have evaluated the volume dependence of the pion mass based on a quark-meson model [26]. Within a lattice regularised ChPT finite volume effects have been addressed by Borasoy et al. in Ref. [27]

The outline of the article is as follows. In sect. 2 we set the notation and remind of the basic assumptions for an application of ChPT in finite volume. Sect. 3 is devoted to outline the calculation and state the main results. In sect. 4 we show the explicit expressions which have been used for the numerical analysis in sect. 5. We conclude with a summary.

2 Preliminaries

In this section we shall set the notation and the basic definitions for the two-loop calculation.

2.1 ChPT in finite and in infinite volume

Chiral Perturbation Theory (ChPT) is the effective theory for QCD at low energies. It is nowadays a mature field which has been applied successfully to a variety of phenomena, in particular in the meson sector. For an introduction and a review of a current status of the field, we refer to Ref. [28, 29, 30, 31].

Let us first restrict ourselves to the infinite volume case. For two flavours the effective Lagrangian of QCD at low energies is made up of an infinite number of terms [2],

$$\mathcal{L}_{\text{eff}} = \mathcal{L}_2 + \mathcal{L}_4 + \mathcal{L}_6 + \ldots . \tag{1}$$

As we wish to calculate the pion mass, an on-shell quantity, external fields can be dropped in $\mathcal{L}_{\text{eff}}$. We work in the isospin symmetry limit $m_u = m_d$ in Euclidean space-time, and for the choice of the pion fields we use the non-linear sigma model parameterisation. We have

$$\mathcal{L}_2 = \frac{F^2}{4} \langle u_\mu u_\mu - \chi_+ \rangle \ , \tag{2}$$

with

$$U = \sigma + i\frac{\phi}{F} \quad , \quad \sigma^2 + \frac{\phi^2}{F^2} = 1 \ ,$$

$$\phi = \begin{pmatrix} \pi^0 & \sqrt{2}\pi^+ \\ \sqrt{2}\pi^- & -\pi^0 \end{pmatrix} = \phi^i \tau^i \ ,$$

$$u_\mu = iu^\dagger \partial_\mu U u^\dagger = -iu \partial_\mu U^\dagger u = u_\mu^\dagger \ ,$$

$$\chi_+ = u^\dagger \chi u^\dagger + u \chi^\dagger u \;,$$

$$\chi = 2B\hat{m}\mathbf{1} \;, \qquad \hat{m} = \frac{1}{2}(m_u + m_d) \;, \tag{3}$$

with $u^2 = U$. The symbol $\langle A \rangle$ denotes the trace of the two–by–two matrix A. The Lagrangian $\mathcal{L}_4$ consists of [2],

$$\mathcal{L}_4 = \sum_{i=1}^{3} \ell_i P_i + \ldots \;, \tag{4}$$

where

$$P_1 = -\frac{1}{4}\langle u_\mu u_\mu \rangle^2 \;, \quad P_2 = -\frac{1}{4}\langle u_\mu u_\nu \rangle \langle u_\mu u_\nu \rangle \;, \quad P_3 = -\frac{1}{16}\langle \chi_+ \rangle^2 \;. \tag{5}$$

The ellipsis in eq. (4) denotes terms that do not contribute to the pion mass. The low energy constants ℓ_i are divergent and remove the ultraviolet divergences generated by one–loop graphs from $\mathcal{L}_2$.

The complete effective Lagrangian $\mathcal{L}_6$ with its divergence structure at $d = 4$ has been constructed in [32, 33]. As will be discussed in sect. 3, terms from $\mathcal{L}_6$ merely renormalize the pion mass in infinite volume and do not contribute to finite volume corrections which we are interested in. Thus, we refrain from showing it here. Given the effective Lagrangian and the parameterisation for the pion fields, it is straightforward to calculate the pion mass to two-loops. We refer to [4], where one also finds a detailed discussion on the renormalization procedure [for technical support on two-loop diagrams in ChPT, see [34]].

The effective framework is still appropriate, when the system is enclosed by a large box of size $V = L^3$. We refer to the literature for the foundations [10, 11, 12] and a recent review [9]. Here, we only remind of the fundamental results which guided the present calculation: the volume has to be large enough, such that ChPT can give reliable results, $2F_\pi L \gg 1$. The perturbative calculation is bound to the value of the parameter $M_\pi L$. Whether it happens to be large ($M_\pi L \gg 1$, "p-regime") or small ($M_\pi L \lesssim 1$, "ϵ-regime") implies a different power counting. However, in both cases the effective Lagrangian is the same as in the infinite volume. In this article we only cover the "p-regime", where the system is distorted mildly and the only change brought about by the finite volume is a modification of the pion propagator due to the periodic boundary conditions of the pion fields[2]

$$G(x^0, \mathbf{x}) = \sum_{\mathbf{n} \in \mathbb{Z}^3} G_0(x^0, \mathbf{x} + \mathbf{n}L) \;, \tag{6}$$

---

[2]Throughout we denote by the volume the three-dimensional volume $V = L^3$, whereas the time direction is not compactified.

with $G_0(x)$ the propagator in infinite volume.

2.2 Basic definitions

In Euclidean space-time the propagator is defined through the connected correlation function

$$G(x)\delta^{ab} = \langle \phi^a(x)\phi^b(0)\rangle_L, \tag{7}$$

where a,b stand for isospin indices of the pion fields and the subscript L in the correlation function denotes that it is evaluated in finite volume. We also have

$$\langle \phi^1(x)\phi^1(0)\rangle_L = L^{-3}\sum_{\mathrm{p}}\int \frac{dp^0}{2\pi}e^{ipx}G(p^0,\mathrm{p}),$$

$$G(p^0,\mathrm{p})^{-1} = M^2 + p^2 - \Sigma_L(p^2), \tag{8}$$

where the momenta p take values

$$\mathrm{p} = \frac{2\pi}{L}\mathrm{n}, \qquad \mathrm{n} \in \mathbb{Z}^3, \tag{9}$$

and $M^2 = 2B\hat{m}$ is the tree-level pion mass in infinite volume. The pion mass in finite volume $M_{\pi L}$ is now defined by the pole equation

$$G(\hat{p}_L)^{-1} = 0, \quad \text{for} \quad \hat{p}_L = (iM_{\pi L}, 0). \tag{10}$$

3 Outline of calculation and statement of results

For a large volume, the finite size effects are expected to be small, such that the pole equation can be solved perturbatively. We outline the calculation and proceed with the main results, whereby derivations are relegated to the upcoming sections.

3.1 One-loop result

Since the effective Lagrangian remains unchanged when going to the finite volume, we can immediately write down the Feynman diagrams which contribute to the self-energy at two-loops, see fig. 1, the only difference with respect to an infinite volume calculation is that the propagators need to be periodified, cf. eq.(6). At

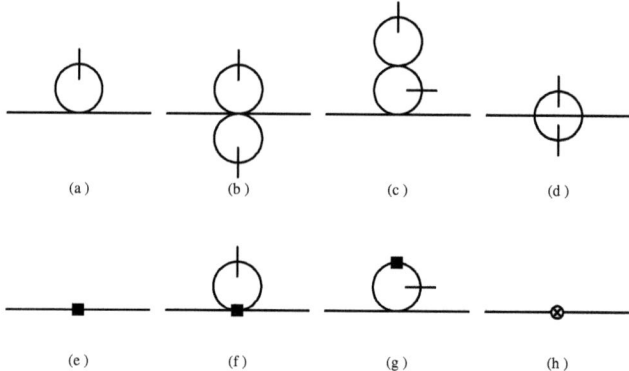

Fig. 1. Self-energy graphs to two-loops in ChPT. A spline corresponds to a periodified propagator, whereas those without correspond to an infinite volume propagator, cf. eq.(6). Normal vertices come from $\mathcal{L}_2$, squared vertices from $\mathcal{L}_4$ and the circle-crossed from $\mathcal{L}_6$.

leading order, the graphs 1(a) and 1(e) need to be evaluated and the self-energy admits the form

$$\Sigma_L(p^2) = \frac{1}{F_\pi^2} G(0) \left(-\frac{3}{2} M_\pi^2 - p^2 \right) - 2\ell_3 \frac{M_\pi^4}{F_\pi^2} + \mathcal{O}\left(\frac{1}{F_\pi^4} \right) , \qquad (11)$$

with $G(0)$ the value of the finite volume propagator at the origin. It contains a logarithmic divergence due to the contribution from the zero mode term n = 0 in eq.(6). In dimensional regularization,

$$G_0(0) = \frac{1}{(2\pi)^d} \int d^d p \frac{1}{p^2 + M^2} = \frac{\Gamma(1 - d/2)}{(4\pi)^{d/2}} M^{d-2} . \qquad (12)$$

The remaining terms with n $\neq$ 0 are finite and may be expressed in terms of a kinematical function $g_1(M^2, 0, L)^3$,

$$G(0) = G_0(0) + g_1(M^2, 0, L),$$
$$g_1(M^2, 0, L) = \int_0^\infty d\tau (4\pi\tau)^{-d/2} e^{-\tau M^2} \sum_{n \neq 0} \exp\left(-\frac{n^2 L^2}{4\tau} \right) \qquad (13)$$

For a derivation of eq.(13), we refer to [35]. In app. A we provide a different derivation, based on a contour integration analysis. The pole of the Gamma function in eq.(12) in four dimensions is absorbed in a renormalization of the low-energy

---
[3] Notation as in Ref.[10]

constant ℓ_3. One readily verifies that inserting eq.(13) into eq.(11) yields for the leading finite volume shift

$$M_{\pi L} = M_\pi \left[1 + \frac{1}{4F_\pi^2} g_1(M_\pi^2, 0, L) + \mathcal{O}\left(\frac{1}{F_\pi^4}\right) \right]. \tag{14}$$

The separation of the cut-off and the volume-dependence is as expected: finite volume corrections do not generate new uv-divergences. At leading order the finite volume corrections could be isolated immediately. This will not be the case at the two-loop level. The graph (d) in fig. 1 does not factorise in pure one-loop integrals and further steps need to be performed. We refer to app. A for further details.

3.2 Minimal set of periodified propagators

In the evaluation of the Feynman diagrams, it is mandatory to make use of the following claim. In a graph consisting of L loops, only a certain set of L propagators have to be periodified. In the following we wish to verify this claim, (see also p.18ff in Ref.[1]). Consider an arbitrary self-energy graph with L-loops, I internal lines ℓ and V vertices. Since the number of loops is the number of independent integrations over momenta, we have

$$L = I - V + 1. \tag{15}$$

We assign for every line ℓ an integer vector $n(\ell)$ and interpret it as a gauge field on the self-energy graph. Gauge transformations are then defined as[4]

$$n'(\ell) = n(\ell) + \Lambda(f(\ell)) - \Lambda(i(\ell)), \tag{16}$$

where $\Lambda(v)$ is some field of integer vectors and $i(\ell)$ ($f(\ell)$) is the initial (final) vertex of the line ℓ. One verifies immediately that two contributions of a Feynman graph which differ only by a gauge transformation are gauge equivalent and yield the same mathematical expression. What matters is the sum over representatives of gauge equivalence classes. A convenient representative can be found by adjusting $\Lambda(i(\ell))$ and $\Lambda(f(\ell))$ iteratively, such that $n(\ell) = 0$ for as many internal lines ℓ as possible. This can be achieved for $V-1$ lines. It thus remain $I-V+1$ internal lines where the periodified propagator has to be inserted – which equals with the number of loops of the graph. This shall be our minimal set of periodified propagators. In fig. 1, we have attached a spline to a periodified propagator, whereas lines without a spline correspond to an infinite volume propagator.

---

[4]Notation as in Ref.[1]

3.3 Two-loop result

It is convenient to split the sum over the equivalence classes into three parts[5]:

$$\begin{aligned}\Sigma^{(0)}: \quad &\text{n}(\ell) = 0 \; \forall \; \ell \; \text{(pure gauge)},\\ \Sigma^{(1)}: \quad &\text{n}(\ell) = 0 \; \forall \; \ell \; \text{except for one line } \bar{\ell} \; \text{(simple gauge)},\\ \Sigma^{(2)}: \quad &\text{n}(\ell) = 0 \; \forall \; \ell \; \text{except for two lines } \bar{\ell}_1, \bar{\ell}_2. \end{aligned} \qquad (17)$$

The pion mass in finite volume to two-loops then admits the form

$$\begin{aligned} M_{\pi L}^2 &= M_\pi^2 - \Sigma^{(1)} - \Sigma^{(2)} \;,\\ M_\pi^2 &= M^2 - \Sigma^{(0)} \;, \end{aligned} \qquad (18)$$

where we find

$$\Sigma^{(1)} = I_p + I_c + \mathcal{O}(\xi^3) \;, \qquad (19)$$

$$\Sigma^{(2)} = M_\pi^2 \xi^2 \left[\frac{9}{8} \tilde{g}_1(\lambda_\pi)^2 - \frac{1}{8} \lambda_\pi \tilde{g}_1(\lambda_\pi) \frac{\partial}{\partial \lambda_\pi} \tilde{g}_1(\lambda_\pi) + \Delta \right] + \mathcal{O}(\xi^3) \;, \qquad (20)$$

$$I_p = \frac{M_\pi^2}{16\pi^2 \lambda_\pi} \sum_{n=1}^\infty \frac{m(n)}{\sqrt{n}} \int_{-\infty}^\infty dy \, \mathcal{F}_\pi(iy) \, e^{-\sqrt{n(1+y^2)}\lambda_\pi} \;, \qquad (21)$$

$$I_c = -\frac{iM_\pi^2}{32\pi^3 \lambda_\pi} \sum_{n=1}^\infty \frac{m(n)}{\sqrt{n}} \int_{-\infty}^\infty dy \int_4^\infty d\tilde{s} \, \frac{e^{-\sqrt{n(\tilde{s}+y^2)}\lambda_\pi}}{\tilde{s}+2iy} \, \text{disc}[\mathcal{F}_\pi(\tilde{s}, 1+iy)] \;, \qquad (22)$$

and introduced the abbreviations

$$\lambda_\pi = M_\pi L \;, \qquad \xi = \frac{M_\pi^2}{16\pi^2 F_\pi^2} \;, \qquad (23)$$

$$m(n) \equiv \text{number of integer vectors z with } z^2 = n.$$

A detailed derivation of these results will be given in the subsequent sections and we confine ourselves to a few comments at this stage: In I_p one recovers the asymptotic formula of Lüscher, if one restricts the sum to the first addend. Its extension to the present form has already been suggested in [9] and was applied in

---

[5]Note the slight difference in the definition of simple fields with respect to [1]. We do not require the restriction $|\text{n}(\ell)| = 1$.

[36]. The function $\mathcal{F}_\pi(\nu)$ denotes the isospin zero $\pi\pi$ scattering amplitude in the forward kinematics. It contains cuts due to the two-pion channel. In the derivation of the asymptotic formula, Lüscher consistently dropped contributions arising from the cuts, since in a large volume expansion they are beyond the order of accuracy he aimed at. Here, we take them into account up to two-loops, relying on the chiral representation of $\mathcal{F}_\pi(\nu)$. The outcome of the analysis is summarized in I_c. The contributions from the cuts can still be written as exponentially weighted integrals over the $\pi\pi$ scattering amplitude, as in the asymptotic formula. Notice however, that this representation relies on its chiral representation and is only valid up to two-loops, whereas the Lüscher formula holds on a non-perturbative level.

Contributions from two pion propagators in finite volume are ultimately captured in $\Sigma^{(2)}$, being expressed in terms of a dimensionless function $\tilde{g}_1(\lambda_\pi)$ and a numerically small correction Δ arising from graph 1 d). Both are explicitly given in app. A and eq. (44), respectively.

3.4 Large L limit

Before we proceed with the derivation of eq.(18–23), we add a remark concerning the large L behaviour of the two-loop results. In the large volume limit, the contributions from eq.(21) behave according to

$$\lim_{L \to \infty} I_p \sim \frac{1}{\lambda_\pi^{3/2}} e^{-\lambda_\pi} \ . \tag{24}$$

Similarly, we may evaluate the large volume behaviour of the terms occurring in eqs.(20,22). It turns out that these are suppressed by at least a factor of $1/\sqrt{\lambda_\pi}$ with respect to eq.(24). We come to the conclusion that the resummed Lüscher formula is still dominating in comparison to other two-loop diagrams in a $1/\sqrt{\lambda_\pi}$ expansion, i.e.

$$\begin{aligned} M_{\pi L} &= \bar{M}_{\pi L}(1 + \mathcal{O}(1/\sqrt{\lambda_\pi})) \ , \\ \bar{M}_{\pi L} &= M_\pi - \frac{1}{2M_\pi} I_p \ . \end{aligned} \tag{25}$$

3.5 Self-energy to 0'th order: $\Sigma^{(0)}$

The pure gauge contributions are not volume dependent and merely renormalize the pion mass, cf. eq. (18). A detailed discussion of this calculation can be found in [37, 4], with which we agree.

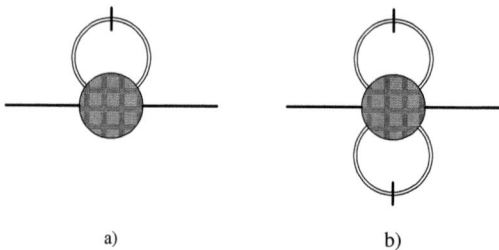

Fig. 2. Skeleton diagrams representing *a)* eq. (26) and *b)* eq. (40). The blob in *a)* stands for the 4-point function of $\pi\pi$ scattering in infinite volume and in *b)* for a subtracted 6-point function of $\pi\pi\pi$ scattering in infinite volume. The double-line with the spline is a finite volume propagator with the physical pion mass M_π^2.

3.6 Self-energy to 1'st order: $\Sigma^{(1)}$

The simple fields can be summed up in closed form and may be displayed in a skeleton diagram, see fig. 2a)

$$\Sigma^{(1)} = \frac{1}{2}\int \frac{d^4q}{(2\pi)^4} \sum_{n=1}^{\infty} m(n)\, e^{iq_1\sqrt{n}L} G_0(q^2)\, \Gamma_{\pi\pi}(\hat{p},q,-\hat{p},-q) \;, \qquad (26)$$

with $\hat{p} = (iM_\pi, 0)$ and $\Gamma_{\pi\pi}(\hat{p},q,-\hat{p},-q)$ the 4-point function of $\pi\pi$ scattering in the forward scattering kinematics[6]. This was already observed by Lüscher and eventually led him to the asymptotic formula [1]. He then discussed the contribution of the pole of the propagator $G_0(q^2)$ that one meets at

$$q_1 = i\sqrt{M_\pi^2 + q_\perp^2 + q_0^2}\,, \qquad q_\perp = (q_2, q_3)\,. \qquad (27)$$

Above this pole the singularities come from the cuts of the propagator and the 4-point function $\Gamma_{\pi\pi}(\hat{p},q,-\hat{p},-q)$. These start from

$$s = -(\hat{p}+q)^2 \geq 4M_\pi^2 \;, \quad u = -(\hat{p}-q)^2 \geq 4M_\pi^2 \;, \quad -q^2 \geq 9M_\pi^2\,. \qquad (28)$$

Lüscher showed that the pole contribution is dominating with respect to those coming from the cuts and neglected the latter. In fact, his discussion involved no further assumptions about the 4-point function and remains to be true at a non-perturbative level. Since our goal is to test the asymptotic formula beyond the leading exponentials, we wish to work out the impact of the contributions that

---

[6] Note that the 4-point function in eq. (26) stands for an off-shell amplitude. We will take up this point later in sect. 3.7

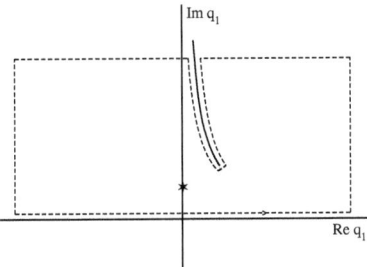

Fig. 3. Integration contour in the complex q_1 plane with the pole from the pion propagator and the branch cut from the $\pi\pi$ scattering amplitude.

were dropped by Lüscher. Due to the involved non-analytic structure of the $\pi\pi$ scattering amplitude at high energies, this task is difficult to be answered at a non-perturbative level. At the two-loop level, these terms may be taken into account with a chiral representation of the scattering amplitude. This shall be addressed now.

Up to $\mathcal{O}(p^4)$, the four-point function can be decomposed into a combination of functions which have either a singularity in s or in u. Since $\Sigma^{(1)}$ is symmetric in s and u, we may write $\Gamma_{\pi\pi}(\hat{p}, q, -\hat{p}, -q)$ in terms of a function $\bar{\Gamma}_{\pi\pi}(s, \hat{p}q)$, which depends only on the variables s and $\hat{p}q$, and whose cuts are lying at $s \geq 4M_\pi^2$,

$$\Gamma_{\pi\pi}(\hat{p}, q, -\hat{p}, -q) = \bar{\Gamma}_{\pi\pi}(s, \hat{p}q) + \mathcal{O}\left(\frac{1}{F_\pi^6}\right). \tag{29}$$

We write $q_1 = x + iy$ and for the domain $s \geq 4M_\pi^2$ we have

$$y^2 - x^2 - q_0^2 - q_\perp^2 \geq 3M_\pi^2, \qquad xy = -q_0 M_\pi. \tag{30}$$

In particular, in the complex q_1 plane we observe that the cut starts above the pole of eq. (27), as illustrated in fig. 3. Performing now the contour integration in the upper half plane we obtain two terms, one of the residuum of the pole I_p and the other from the integral along the new integration path. The latter contribution vanishes as we push the integration lines to infinity, except for the one along the cut, to be denoted by I_c in the following,

$$\Sigma^{(1)} = I_p + I_c. \tag{31}$$

Along the lines discussed in [1], the former can be simplified considerably and leads to the asymptotic formula

$$I_p = \frac{M_\pi^2}{16\pi^2 \lambda_\pi} \sum_{n=1}^{\infty} \frac{m(n)}{\sqrt{n}} \int_{-\infty}^{\infty} dy\, \mathcal{F}_\pi(iy)\, e^{-\sqrt{n(1+y^2)}\lambda_\pi}\,, \tag{32}$$

with $\mathcal{F}_\pi(iy)$ the $\pi\pi$ forward scattering amplitude. Restricting the sum to the first addend, we recover Lüscher's formula [1]. For I_c we find

$$I_c = \frac{1}{2} \sum_{n=1}^{\infty} m(n) \int \frac{dq_0 d^2q_\perp}{(2\pi)^4} \int_{[s \geq 4M_\pi^2]} dq_1 \frac{e^{iq_1\sqrt{n}L}}{M_\pi^2 + q^2}\, \text{disc}[\bar{\Gamma}_{\pi\pi}(s,\hat{p}q)] + \mathcal{O}(F_\pi^{-6})\,, \tag{33}$$

where $\text{disc}[\bar{\Gamma}_{\pi\pi}(s,\hat{p}q)]$ denotes the discontinuity of $\bar{\Gamma}_{\pi\pi}$ along the cut. It is now convenient to shift the integration path in q_0 from $\Im(q_0) = 0$ to $\Im(q_0) = -iM_\pi$. Along this path we have

$$q_0 = \bar{q}_0 - iM_\pi\,, \quad s = -\bar{q}_0^2 - q_1^2 - q_\perp^2\,, \quad \bar{q}_0 \in \mathbb{R}\,, \tag{34}$$

and the integration over $q_1 = x + iy$ falls onto the imaginary axses,

$$x = 0\,, \quad y \geq y_0 = \sqrt{4M_\pi^2 + \bar{q}_0^2 + q_\perp^2}\,,$$

$$I_c = \frac{i}{2} \sum_{n=1}^{\infty} m(n) \int \frac{d\bar{q}_0 d^2q_\perp}{(2\pi)^4} \int_{y_0}^{\infty} dy\, e^{-y\sqrt{n}L} \frac{\text{disc}[\bar{\Gamma}_{\pi\pi}(s,\hat{p}q)]}{M_\pi^2 + q^2}\,. \tag{35}$$

Next, we change the integration variable from q_1 to s,

$$I_c = -\frac{i}{2} \sum_{n=1}^{\infty} m(n) \int \frac{d\bar{q}_0 d^2q_\perp}{(2\pi)^4} \int_{4M_\pi^2}^{\infty} ds\, \frac{e^{-\sqrt{n(s+\bar{q}_0^2+q_\perp^2)}L}}{2(s+\bar{q}_0^2+q_\perp^2)^{1/2}} \frac{\text{disc}[\bar{\Gamma}_{\pi\pi}(s,\hat{p}q)]}{s + 2iM_\pi \bar{q}_0}\,, \tag{36}$$

and make use of

$$\int \frac{d^2q_\perp}{(2\pi)^2} \frac{1}{2(\mu^2+q_\perp^2)^{1/2}} e^{-\sqrt{n(\mu^2+q_\perp^2)}L} = \frac{1}{4\pi\sqrt{n}L} e^{-\mu\sqrt{n}L}\,, \tag{37}$$

to carry through the integration over $q_\perp$ and to end up with eq. (22).

3.7 Self-energy to 2'nd order: $\Sigma^{(2)}$

In view of the results of the preceding section, the question arises, whether the self-energy to 2'nd order can be represented in a similar compact form as in the case of the self-energy to 1'st order in eq. (26). As will be discussed, it is indeed possible

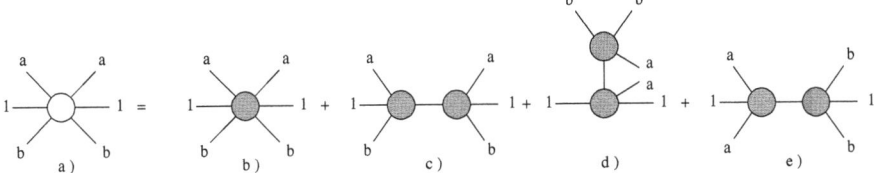

Fig. 4. Decomposition of the 3π-3π amplitude in terms of 1particle irreducible parts. The characters 1, a and b on the external legs denote isospin indices.

to relate the self-energy to 2'nd order to a 3π scattering amplitude in the forward scattering kinematics, as illustrated in fig. 2b). This has already been observed by Schenk in a related context [20]. He investigated the dynamics of pions in a cold heat bath of temperature T and examined the effective mass of the pion $M_\pi(T)$ within the framework of ChPT. The close relation between finite temperature field theory and finite volume effects becomes apparent in the imaginary time formalism, where one treats the inverse temperature as a finite extension in the imaginary time direction. The expansion in terms of the number of finite volume propagators is then in one-to-one correspondence with the expansion in terms of the number of finite temperature propagators. In the following, Schenk's approach shall be adapted to the finite volume scenario. We first establish the relation between $\Sigma^{(2)}$ and the three-to-three particle scattering amplitude and proceed with various remarks.

Consider the 6-point function in d dimensions in the non-linear sigma model parameterisation eq. (3) in the forward kinematics

$$\sum_{a,b=1}^{3} \int dx_1 \ldots dx_5 e^{-ip(x_1-x_4)-ik(x_2-x_5)-iq(x_3-x_6)} \langle 0|T\varphi^1_{x_1}\varphi^a_{x_2}\varphi^b_{x_3}\varphi^1_{x_4}\varphi^a_{x_5}\varphi^b_{x_6}|0\rangle$$

$$= \frac{Z^3}{(M_\pi^2+p^2)^2(M_\pi^2+k^2)^2(M_\pi^2+q^2)^2} T_{\pi\pi\pi}(p,k,q) \; , \tag{38}$$

with Z the wave function renormalization constant and $\varphi_x \equiv \varphi(x)$. The amplitude $T_{\pi\pi\pi}(p,k,q)$ contains a pole at $\hat{p}^2 = -M_\pi^2$ which needs to be subtracted

$$T_{\pi\pi\pi}(p,k,q) = \hat{T}_{\pi\pi\pi}(p,k,q) + \frac{R(p,k,q)}{M_\pi^2+p^2} \; . \tag{39}$$

The self-energy to 2'nd order can then be written in terms of the subtracted amplitude $\hat{T}_{\pi\pi\pi}(\hat{p},k,q)$

$$\Sigma^{(2)} = \frac{1}{8} \sum_{n,r}{}' \int \frac{d^4q}{(2\pi)^4} \frac{d^4k}{(2\pi)^4} e^{iqnL+ikrL} \frac{\hat{T}_{\pi\pi\pi}(\hat{p}, k, q)}{(M_\pi^2 + q^2)(M_\pi^2 + k^2)} + \mathcal{O}(\frac{1}{F_\pi^6}) \ . \quad (40)$$

The meaning of the prime in the sum is postponed to the end of the comments.

i) The reason for the subtraction is easily accounted for. In fig. 4 we decompose the 3π-3π scattering amplitude in terms of 1-particle irreducible parts. The subtraction aims at removing part fig. 4e) which does not correspond to a 1-particle irreducible self-energy diagram, once its external legs are glued together.

A thorough discussion on the physical interpretation of the pole term can be found in [20].

ii) Notice that eq. (39) defines a subtracted off-shell amplitude which depends on the regularization scheme as well as on the parameterisation of the pion fields. At the order we are working, the regularization dependence is not an issue, since the subtracted amplitude $\hat{T}_{\pi\pi\pi}(\hat{p}, k, q)$ is only needed at tree level. However, the dependence on the parameterisation of the pion fields is of concern. While the momentum integrations in eq. (40) for the diagrams fig. 4b) and 4d) put the momenta k and q on-shell and the ambiguity on the parameterisation therefore drops out, it does not in the case of diagram 4c).

Note that the same parameterisation ambiguity already occurred in I_c in the dispersive analysis of $\Sigma^{(1)}$ (cf. eq. (34) where $q^2 \neq -M_\pi^2$). In order to understand the close relation between these two terms, we first note that only diagram fig. 1(d) contributed to I_c, which corresponds to fig. 4c). Further, the simple gauge field of fig. 1(d) with n $\neq$ 0 for one propagator can immediately be written as a contribution with two finite volume propagators: one periodifies a second propagator, however with the same n $\neq$ 0 as already for the first one. Since $M_{\pi L}$ does not depend on the parameterisation of the pion fields, the ambiguities of the two terms have to cancel each other. To see this algebraically is however by no means trivial.

In summary, while our representations for I_c in eq. (22), resp. eq. (41) and for $\Sigma^{(2)}$ in eq. (40), resp. eq. (20) do depend on the off-shell dependence of the scattering amplitudes, the sum in eq. (18) does not.

iii) Even though the self-energy to 2'nd order can be expressed in a compact form, further simplifications (similar to those performed in sect. 3.6) seem not to be straightforward. Instead, for the (numerical) integrations we had to discuss the various terms contributing to $\Sigma^{(2)}$ one by one. App. A is devoted to finite volume integrals and shall discuss the main steps to be performed. After trading the finite volume integrals, we end up with the basic functions given in eq. (20).

iv) The prime in eq. (40) restricts the sum to integer vectors n and r obeying $n \neq 0 \neq r$, and for diagram c) in fig. 4 in addition $n \neq r$. The latter restriction preserves double counting as this term is already accounted for in a simple gauge field of $\Sigma^{(1)}$.

4 Summary of analytical results

In this section we shall summarise the analytical results and express them in terms of a few basic integrals. In eq.(18-23) we have split the finite size effects of the pion mass into I_p, I_c and $\Sigma^{(2)}$. For the former two we find

$$I_p = M_\pi^2 \sum_{n=1}^\infty \frac{m(n)}{\sqrt{n}} \frac{1}{\lambda_\pi} \xi \left[I_{M_\pi}^{(2)} + \xi I_{M_\pi}^{(4)} + \xi^2 I_{M_\pi}^{(6)} \right] ,$$

$$I_c = M_\pi^2 \sum_{n=1}^\infty \frac{m(n)}{\sqrt{n}} \frac{1}{\lambda_\pi} \xi^2 \tilde{I}_c , \qquad (41)$$

where ξ is the chiral expansion parameter defined in eq.(23) and the expressions $I_{M_\pi}^{(2)}$, $I_{M_\pi}^{(4)}$ and $I_{M_\pi}^{(6)}$ have already been introduced in Ref.[36]. The coefficient $\tilde{I}_c$ can be expressed as a combination of basic integrals

$$\tilde{I}_c = \frac{1}{3} \left(112 C^{0,2} + 37 C^{1,0} - 40 C^{1,2} - 4 C^{2,0} \right) ,$$

$$C^{j,k} = \int_{-\infty}^\infty dy \int_4^\infty d\tilde{s} \, \frac{e^{-\sqrt{n(\tilde{s}+y^2)}\lambda_\pi}}{\tilde{s}^2 + 4y^2} \left(1 - \frac{4}{\tilde{s}}\right)^{1/2} \tilde{s}^j y^k . \qquad (42)$$

Eventually, the self-energy to 2'nd order has already been introduced in eq.(20)

$$\Sigma^{(2)} = M_\pi^2 \xi^2 \left[\frac{9}{8} \tilde{g}_1(\lambda_\pi)^2 - \frac{1}{8} \lambda_\pi \tilde{g}_1(\lambda_\pi) \frac{\partial}{\partial \lambda_\pi} \tilde{g}_1(\lambda_\pi) + \Delta \right] + \mathcal{O}(\xi^3) , \qquad (43)$$

with

$$\Delta = (16\pi^2)^2 \left[4 \tilde{p}_\mu \tilde{p}_\nu H_{\mu\nu} + 4 \tilde{p}_\mu H_\mu + \frac{7}{6} H \right] , \qquad \tilde{p} = \frac{\hat{p}}{M_\pi} , \qquad (44)$$

where H, H_μ and $H_{\mu\nu}$ are related to the sunset-type integrals of fig.1d). We have not been able to find a compact representation for these integrals and only elaborate on their numerical analysis in app. A. in some detail.

Note that while the pole term I_p has been evaluated up to order ξ^3, the contributions from the contour I_c and those from two-propagators in finite volume $\Sigma^{(2)}$ are only known to order ξ^2. It would be very difficult to draw level with I_c and $\Sigma^{(2)}$ to the same order. The reason why it is still not questionable to keep I_p to order ξ^3 is due to its exponential leading large L contributions. The numerical analysis in the next section confirms the expectation that whenever we are in the p-regime, the corrections from I_c and $\Sigma^{(2)}$ are negligible compared to the asymptotic contributions of I_p. A calculation to order ξ^3 for these terms is therefore not needed.

5 Numerics

5.1 Setup

The numerical analysis is performed in line with the setup of Ref.[36]. Therefore, we shall be rather short in the following. The quantity of interest is

$$R_{M_\pi} \equiv \frac{M_{\pi L} - M_\pi}{M_\pi}, \tag{45}$$

whose quark mass dependence shall be evaluated numerically for different sizes L. The parameters of R_{M_π} are (see eqns.(18–23) and eqns.(41-44)) the pion mass M_π and the pion decay constant F_π in infinite volume as well as (implicitly in $I_{M_\pi}^{(4/6)}$) the SU(2) low energy constants. The quark mass dependence of the pion decay constant may be taken into account by expressing F_π as a function of the pion mass M_π. Regarding the low energy constants, we use the ones determined in [4, 5] which are the same as in our previous finite size studies [7, 36].

5.2 Results

We plot our results for R_{M_π} in fig. 5, both for $L = 2, 3, 4$fm as a function of M_π and for $M_\pi = 100, 300, 500$MeV as a function of L. We show the one-loop result (LO) as well as the two-loop result (NLO). These shall be compared with the resummed asymptotic formula with LO/NLO/NNLO input for the $\pi\pi$ scattering amplitude. Note that the one-loop result and the resummed asymptotic formula to LO coincide. The best estimate for R_{M_π} is finally obtained by adding to the asymptotic pure three-loop contribution the two-loop result (NNLO asympt. + NLO non-asympt.). At NLO, the finite size effects contain low energy constants, see eg. diagram f) and g) in fig. 1, leading to a non-negligible error band.

We take up a point which was already observed in [6], namely the large contributions when going from LO to NLO in the asymptotic formula (dotted to thin-dash-dotted). Compared to this gap, the additional contributions from the full two-loop result (thick-dash-dotted) are very small. The two-loop and the NLO result from the asymptotic formula only drift away, when we go beyond the region where the p-regime can be safely applied. In tab. 1 we wish to underline these statements with a numerical example: we show the relative numerical impact of the LO, the pure NLO and the pure NNLO contribution for R_{M_π}. In the second column we give the source of the effect. Eg. the fifth line is to be read as the solely contribution of the dispersive terms I_c for R_{M_π}, once kinematical prefactors are added accordingly. We observe that I_c is strongly suppressed, irrespective of the value $M_\pi L$. Consider the column with $M_\pi L = 1.4$. Although the main bulk of the subleading effects is still due to the asymptotic contributions, the terms of $\Sigma^{(2)}$ play a significant role and can not be neglected. This behaviour was expected, since with $M_\pi L = 1.4$ we might have already crossed the border of the p-regime. As we increase this parameter to $M_\pi L = 2$, the asymptotic regime begins to set in. The contributions from the resummed asymptotic formula are now dominating with respect to those from $\Sigma^{(2)}$. The numerical results for $M_\pi L = 2.5$ confirm this trend. The fact that even the additional NNLO asymptotic terms (a partial three-loop result) are lager than the NLO non-asymptotic contributions is in nice agreement with the analytical expectation found in eq. (25). However, we find it disputatious whether the argument given is the main source for the rather enhanced suppression or not. Finally, the discussion of the subleading effects allows us to give a reliable estimate for the lower bound of $M_\pi L$ for the p-regime,

$$M_\pi L \gtrsim 2 : \quad \text{lower bound for } p\text{-regime} . \qquad (46)$$

R_{M_π} $L = 2$ fm			$M_\pi = 140$ MeV $M_\pi L = 1.4$	$M_\pi = 200$ MeV $M_\pi L = 2.0$	$M_\pi = 250$ MeV $M_\pi L = 2.5$
	I_p	LO	0.0453	0.0198	0.0104
	I_p	ΔNLO	0.0275	0.0145	0.0087
	I_c	ΔNLO	0.0005	0.0001	0.0000
	$\Sigma^{(2)}$	ΔNLO	0.0186	0.0038	0.0011
	I_p	ΔNNLO	0.0071	0.0052	0.0035
R_{M_π}			0.0988(49)	0.0434(48)	0.0237(40)

Tab. 1. $R_{M_\pi} = M_{\pi L}/M_\pi - 1$ for a 2 fm volume and pion masses $M_\pi = 140$MeV, 200MeV, 250MeV. We show the relative numerical impact of the LO, the pure NLO and the pure NNLO contribution for R_{M_π}. In the second column we give the source of the effect. Eg. the fifth line is to be read as the solely contribution of the dispersive terms I_c for R_{M_π}, once kinematical prefactors are added accordingly.

6 Summary

i) We have evaluated the finite volume corrections for the pion mass to two-loops within the framework of Chiral perturbation theory (ChPT) in the p-regime ($M_\pi L \gg 1, L > 2$fm, $M_\pi < 500$MeV).

ii) The analytical results are compared with an extended version of the asymptotic formula of Lüscher. We have found that whenever the effects are calculated in the proper p-regime of ChPT, the contributions which are not included in the asymptotic formula are very small.

iii The derivation of the asymptotic formula of Lüscher does not rely on the specific interactions among the particles under consideration – its asymptotic behaviour is universal. We expect therefore that the asymptotic formula for decay constants holds on a similar level of accuracy.

iv) The discussion about the subleading finite volume effects allows us to estimate a reliable lower bound for the p-regime: $M_\pi L \gtrsim 2$ in the case of the pion mass.

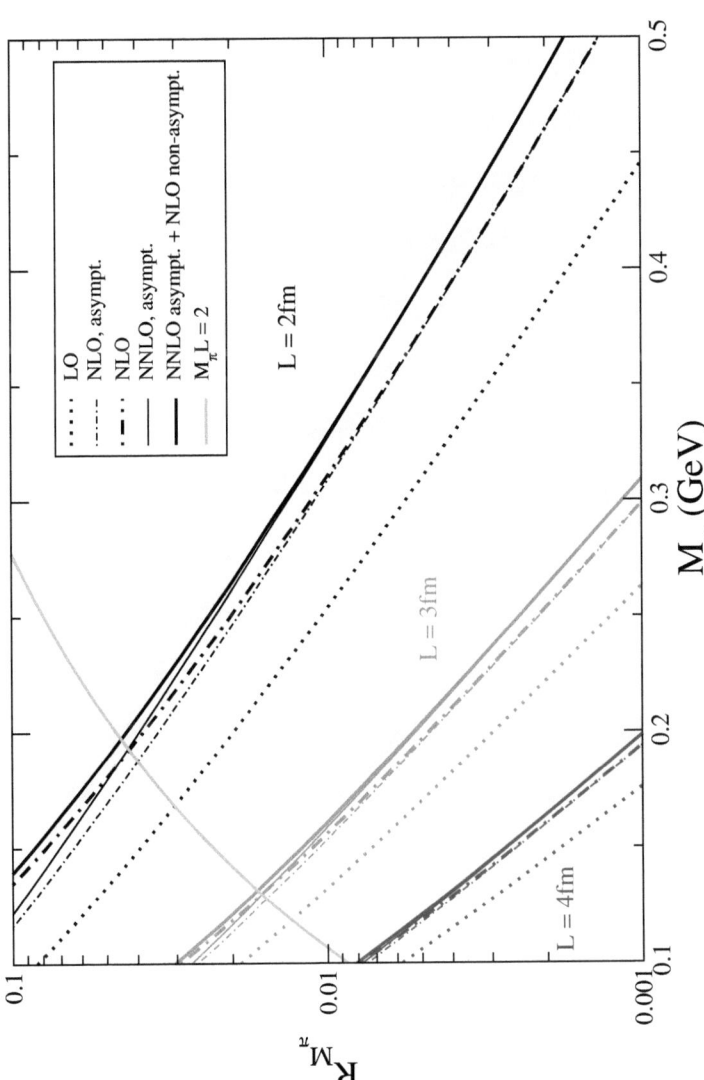

Fig. 5. $R_{M_\pi} = M_{\pi L}/M_\pi - 1$ vs. M_π for $L = 2, 3, 4$fm. The result of the resummed asymptotic Lüscher formula (21) with LO/NLO/NNLO chiral input (with attribute "asympt." for NLO and NNLO in legend) is compared to the one-loop (LO) and two-loop (NLO) result. The best estimate for R_{M_π} is obtained by adding the pure three loop contribution from the asymptotic formula to the two-loop result (NNLO asympt.+ NLO non-asympt.). The low energy constants lead to a non-negligible error band which is only shown for the best estimate. In the region above the $M_\pi L = 2$ line, one is not safely in the p-regime and our results should not be trusted.

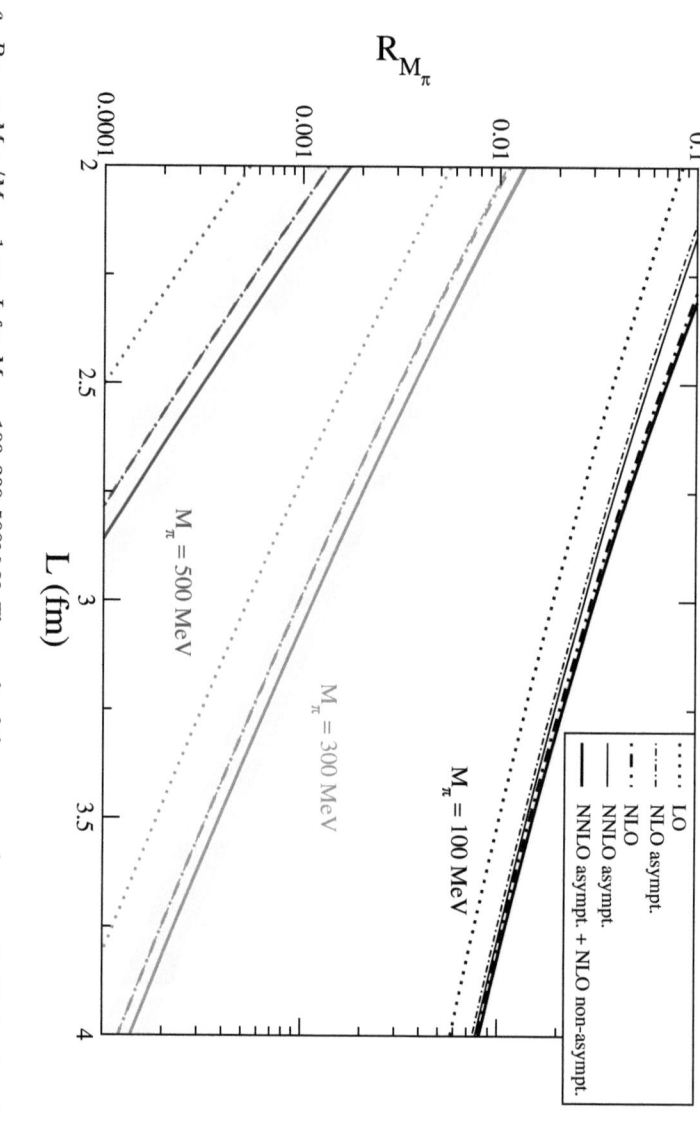

Fig. 6. $R_{M_\pi} = M_{\pi L}/M_\pi - 1$ vs. L for $M_\pi = 100, 300, 500$ MeV. The result of the resummed asymptotic Lüscher formula (21) with LO/NLO/NNLO chiral input (with attribute "asympt." for NLO and NNLO in legend) is compared to the one-loop (LO) and two-loop (NLO) result. The best estimate for R_{M_π} is obtained by adding the pure three loop contribution from the asymptotic formula to the two-loop result (NNLO asympt.+ NLO non-asympt.). The low energy constants lead to a non-negligible error band which is only shown for the best estimate. In the region above $M_\pi L = 2$, one is not safely in the p-regime and our results should not be trusted.

A Finite volume integrals

In this appendix we give further details on the finite volume integrals which occurred in the two-loop calculation. Throughout we applied dimensional regularization, as is common in ChPT. Since the finite box breaks Lorentz invariance, tensor simplifications have to be performed with care. Consider eg. the integral

$$\hat{p}_\mu \hat{p}_\nu A_{\mu\nu} = \int \frac{d^d q}{(2\pi)^d} \sum_{n \in \mathbb{Z}^3}{}' \frac{e^{iqnL}}{M_\pi^2 + q^2} (\hat{p}q)^2 \ , \qquad (47)$$

which arises when evaluating diagram $f)$ in fig. 1. Remember $\hat{p} = (iM_\pi, 0)$. In infinite volume the ansatz

$$A_{\mu\nu} = \delta_{\mu\nu} A \qquad (48)$$

with A a scalar integral allows one to boil down the tensor integral to a well known scalar integral. If we performed the same steps in finite volume, we would obtain

$$\hat{p}_\mu \hat{p}_\nu A_{\mu\nu} = \frac{(M_\pi^2)^{1+d/2}}{d} \sum_{n=1}^{\infty} \frac{m_d(n)}{n^{(d-2)/4}} \frac{1}{(2\pi)^{d/2}} \frac{1}{\lambda_\pi^{(d-2)/2}} K_{(d-2)/2}(\lambda_\pi \sqrt{n}) \ , \qquad (49)$$

with $K_{(d-2)/2}(x)$ a modified Bessel function and $m_d(n)$ defined similarly to eq.(23), but in d dimensions. However, the ansatz (48) is not justified in finite volume. A direct calculation of the integral with the help of eq. (62) yields

$$\hat{p}_\mu \hat{p}_\nu A_{\mu\nu} = -\frac{(M_\pi^2)^{1+d/2}}{(2\pi \lambda_\pi)^{d/2}} \sum_{n=1}^{\infty} \frac{m_d(n)}{n^{d/4}} K_{d/2}(\lambda_\pi \sqrt{n}) \ . \qquad (50)$$

In four dimensions the result agrees with the derivation outlined in the next section, where one proceeds along the same lines as in sect. 3.6.

A.1 Tadpole

In fig. 1, the finite volume corrections of the diagrams (a)–$(c),(f)$ and (g) factorise into one-loop integrals which admit the generic form

$$\int \frac{d^4 q}{(2\pi)^4} \sum_{n \in \mathbb{Z}^3}{}' \frac{e^{iqnL}}{(M_\pi^2 + q^2)^k} \mathcal{P}(\hat{p}q; q^2) \ , \qquad (51)$$

with k an integer positive number and $\mathcal{P}(\hat{p}q, q^2)$ a polynomial of $\hat{p}q$ and q^2. The prime in the sum denotes that the term with n = 0 is excluded. It suffices to discuss the case for $k = 1$, since $k = 2, 3, \ldots$ are obtained through appropriate derivatives with respect to M_π^2. The only singularity of the function under the integral is met by the pole of the propagator. Therefore, the contour integration analysis applied in sect. 3.6 yields the result

$$\sum_{n=1}^{\infty} \frac{m(n)}{\sqrt{n}} \frac{M_\pi^2}{8\pi^2 \lambda_\pi} \int_{-\infty}^{\infty} dy \, e^{-\sqrt{n(1+y^2)}\lambda_\pi} \mathcal{P}(iM_\pi^2 y; -M_\pi^2) \; . \tag{52}$$

In particular, in the text we have used the dimensionless function $\tilde{g}_1(\lambda_\pi)$,

$$\tilde{g}_1(\lambda_\pi) = \frac{16\pi^2}{M_\pi^2} g_1(M_\pi^2, 0, L) \; ,$$

$$g_k(M_\pi^2, 0, L) = \sum_{n \in \mathbb{Z}^3}{}' \int \frac{d^d q}{(2\pi)^d} \frac{e^{iqnL}}{(M_\pi^2 + q^2)^k} \; , \tag{53}$$

evaluated at $d = 4$, and where $g_1(M_\pi^2, 0, L)$ was introduced a long time ago by Gasser and Leutwyler [10, 11], see also eq. (13).

A.2 Sunset

The only real two-loop diagram is the sunset graph fig. 1d), and we will comment on it in some detail. It is convenient to split the finite volume integrals as in eq.(17) only after the tensor simplifications. As alluded in the beginning of the appendix, tensor simplifications in finite volume have to be performed with care. Even though one can not rely on Lorentz invariance, the sunset tensor integrals may still be reduced to the structures

$$\{\mathcal{H}, \mathcal{H}_\mu, \mathcal{H}_{\mu\nu}\} = \int d^d x \, e^{ipx} G(x)^2 \Big[\{1, i\partial_\mu, -\partial_\mu \partial_\nu\} G(x)\Big] \; , \tag{54}$$

with $G(x)$ from eq. (6). A rather direct way to perform these steps is to work in coordinate space and to make use of partial integrations, i.e.

$$\int d^d x \, e^{ipx} \partial_\mu G(x) \partial_\nu G(x) G(x) =$$

$$-\frac{1}{2} \int d^d x \, e^{ipx} G(x)^2 \Big[p_\mu i\partial_\nu G(x) + \partial_\mu \partial_\nu G(x)\Big] \; . \tag{55}$$

Notice that the same identities in momentum space are derived with the help of translational invariance which is still respected in finite volume due to the periodic boundary conditions. In the following, we will only elaborate on the scalar integral $\mathcal{H}$ in more detail. We expand the integral in terms of number of finite volume propagators as motivated in sect. 3.3,

$$\mathcal{H} = \mathcal{H}^{(0)} + 3\mathcal{H}^{(1)} + \mathcal{H}^{(2)} , \qquad (56)$$

where the first (second) addend corresponds to the pure (simple) gauge fields contribution. For $\mathcal{H}^{(0)}$ we refer to [34]. Further,

$$\mathcal{H}^{(1)} = \sum_{n=1}^{\infty} m(n) \int \frac{d^4q}{(2\pi)^4} \frac{e^{iq_1\sqrt{n}L}}{M_\pi^2 + q^2} \bar{J}(\hat{p}-q) + g_1(M_\pi^2, 0, L) J(0) , \qquad (57)$$

with

$$\begin{aligned} J(k) &= \bar{J}(k) + J(0) \\ &= \int \frac{d^d\ell}{(2\pi)^d} \frac{1}{[M_\pi^2 + \ell^2]} \frac{1}{[M_\pi^2 + (k-\ell)^2]} . \end{aligned} \qquad (58)$$

The first term of eq. (57) is finite, the second carries an uv-divergence which is absorbed by a counterterm. This shows that although finite volume effects do not generate new uv-divergences, they still appear at intermediate steps of the calculation. It is a thorough check on our calculation that these non-analytic divergences cancel. Finally,

$$\mathcal{H}^{(2)} = \sum_{\substack{n,r\in\mathbb{Z}^3\backslash 0 \\ n\neq r}} \int \frac{d^4q}{(2\pi)^4} \frac{d^4\ell}{(2\pi)^4} \frac{e^{iqnL}}{[M_\pi^2+q^2]} \frac{e^{i\ell rL}}{[M_\pi^2+\ell^2]} \frac{1}{[M_\pi^2+(\hat{p}-q-\ell)^2]} . \qquad (59)$$

In the text, we used its dimensionless version

$$\{H; H_\mu; H_{\mu\nu}\} = \sum_{\substack{n,r\in\mathbb{Z}^3\backslash 0 \\ n\neq r}} \int \frac{d^4\tilde{q}}{(2\pi)^4} \frac{d^4\tilde{\ell}}{(2\pi)^4} \frac{e^{i\tilde{q}n\lambda_\pi}}{[1+\tilde{q}^2]} \frac{e^{i\tilde{\ell}r\lambda_\pi}}{[1+\tilde{\ell}^2]} \frac{\{1; \tilde{\ell}_\mu; \tilde{\ell}_\mu\tilde{\ell}_\nu\}}{[1+(\tilde{p}-\tilde{q}-\tilde{\ell})^2]} , \qquad (60)$$

which are uv-finite and need not to be renormalized. Therefore, it only remains to find a representation for these terms which is feasible for a numerical analysis. We shall restrict ourselves to the scalar integral in the following. Introducing a Feynman parameter by combining the second and third denominator, we find

$$H = \int \frac{d^4\tilde{q}}{(2\pi)^4} \frac{d^4\tilde{\ell}}{(2\pi)^4} \int_0^1 dx \sum_{\substack{n,r\in\mathbb{Z}^3\backslash 0 \\ n\neq r}} \frac{1}{[1+\tilde{q}^2]} \frac{e^{i\tilde{q}(r-nx)\lambda_\pi + in\tilde{\ell}\lambda_\pi}}{[1+(\tilde{p}-\tilde{q})^2 x(1-x) + \tilde{\ell}^2]^2} . \qquad (61)$$

We use Schwinger's trick for both remaining denominators

$$\frac{1}{1+x^2} = \int_0^\infty d\alpha \, e^{-\alpha(1+x^2)} \,. \tag{62}$$

The integrals over $\tilde{\ell}$ and $\tilde{q}$ are then of the Gaussian type and can be performed analytically. We are then left with three integrations over a rather lengthy expression which shall not be written down here. Despite their unhandy form, the integrations may still safely be performed numerically. The accuracy of the determination of eq.(61) is not limited by the integration routine, but by the rather slow convergence of the sum in n and r for moderate λ_π. Consequently, the evaluation of the sunset integrals going into Δ is restricted to three significant digits. (The last digit given for $\Sigma^{(2)}$ in tab. 1 is not significant.) Note that the uncertainty of the H-type integrals is not a serious matter. Firstly, it could be lowered by brute force and secondly it only plays a (minor) role, in case when the p-regime can not be safely applied anymore.

References

[1] M. Luscher, Commun. Math. Phys. 104, 177 (1986).

[2] J. Gasser and H. Leutwyler, Annals Phys. 158, 142 (1984).

[3] J. Bijnens, G. Colangelo, G. Ecker, J. Gasser and M. E. Sainio, Phys. Lett. B 374, 210 (1996) [hep-ph/9511397].

[4] J. Bijnens, G. Colangelo, G. Ecker, J. Gasser and M. E. Sainio, Nucl. Phys. B 508, 263 (1997) [Erratum-ibid. B 517, 639 (1998)] [hep-ph/9707291].

[5] G. Colangelo, J. Gasser and H. Leutwyler, Nucl. Phys. B 603, 125 (2001) [hep-ph/0103088].

[6] G. Colangelo, S. Durr and R. Sommer, Nucl. Phys. Proc. Suppl. 119, 254 (2003) [hep-lat/0209110].

[7] G. Colangelo and S. Durr, Eur. Phys. J. C 33, 543 (2004) [hep-lat/0311023].

[8] G. Colangelo and C. Haefeli, Phys. Lett. B 590, 258 (2004) [hep-lat/0403025].

[9] G. Colangelo, Nucl. Phys. Proc. Suppl. 140, 120 (2005) [hep-lat/0409111].

[10] J. Gasser and H. Leutwyler, Phys. Lett. B 184, 83 (1987).

[11] J. Gasser and H. Leutwyler, Phys. Lett. B 188, 477 (1987).

[12] J. Gasser and H. Leutwyler, Nucl. Phys. B 307, 763 (1988).

[13] S. R. Sharpe, Phys. Rev. D 46, 3146 (1992) [hep-lat/9205020].

[14] A. Ali Khan et al. [QCDSF-UKQCD Collaboration], Nucl. Phys. B 689, 175 (2004) [hep-lat/0312030].

[15] S. R. Beane, Phys. Rev. D 70, 034507 (2004) [hep-lat/0403015].

[16] S. R. Beane and M. J. Savage, Phys. Rev. D 70, 074029 (2004) [hep-ph/0404131].

[17] D. Becirevic and G. Villadoro, Phys. Rev. D 69, 054010 (2004) [hep-lat/0311028].

[18] D. Arndt and C. J. D. Lin, Phys. Rev. D 70, 014503 (2004) [hep-lat/0403012].

[19] J. Bijnens, N. Danielsson, K. Ghorbani and T. Lahde, [hep-lat/0509042].

[20] A. Schenk, Phys. Rev. D 47, 5138 (1993).

[21] D. Toublan, Phys. Rev. D 56, 5629 (1997) [hep-ph/9706273].

[22] C. Haefeli, [hep-lat/0509078].

[23] M. Luscher, DESY 83/116 Lecture given at Cargese Summer Inst., Cargese, France, Sep 1-15, 1983

[24] Y. Koma and M. Koma, Nucl. Phys. B 713, 575 (2005) [hep-lat/0406034]. Y. Koma and M. Koma, [hep-lat/0504009].

[25] A. Fuhrer, The nucleon in finite volume, Master Thesis, Universität Bern (2004). Can be obtained from http://www.itp.unibe.ch/index.html?lang=0&id=2&subsubid=0 . G. Colangelo and A. Fuhrer, to be published.

[26] J. Braun, B. Klein and H. J. Pirner, Phys. Rev. D 71, 014032 (2005) [hep-ph/0408116].

[27] B. Borasoy, G. M. von Hippel, H. Krebs and R. Lewis, [hep-lat/0509007].

[28] J. Gasser, [hep-ph/0312367].

[29] S. Scherer and M. R. Schindler, [hep-ph/0505265].

[30] G. Colangelo and G. Isidori, [hep-ph/0101264].

[31] U. G. Meissner, H. W. Hammer and A. Wirzba, [hep-ph/0311212].

[32] J. Bijnens, G. Colangelo and G. Ecker, JHEP 9902, 020 (1999) [hep-ph/9902437].

[33] J. Bijnens, G. Colangelo and G. Ecker, Annals Phys. 280, 100 (2000) [hep-ph/9907333].

[34] J. Gasser and M. E. Sainio, Eur. Phys. J. C 6, 297 (1999) [hep-ph/9803251].

[35] P. Hasenfratz and H. Leutwyler, Nucl. Phys. B 343, 241 (1990).

[36] G. Colangelo, S. Durr and C. Haefeli, Nucl. Phys. B 721, 136 (2005) [hep-lat/0503014].

[37] U. Burgi, Nucl. Phys. B 479, 392 (1996) [hep-ph/9602429].

IV

The pion mass in finite volume to two loops

published in

Proc. of XXIII Int. Symp. on Lattice Field Theory

Pos(LAT2005)059

The pion mass in finite volume to two loops

Christoph Haefeli

Institut für Theoretische Physik, Universität Bern
Sidlerstr. 5, 3012 Bern, Switzerland

Abstract

We evaluate the pion mass in finite volume to two loops within Chiral Perturbation Theory. The results are compared with a recently proposed extension of the asymptotic formula of Lüscher. We find that contributions, which were neglected in the latter, are numerically very small at the two-loop level.

1 Introduction

Numerical simulations with lattice QCD are bound to rather small lattice volumes when determining the hadron spectrum and other low energy parameters in QCD. The computed observables show a volume dependence and a thorough understanding of these effects is important for a correct interpretation of numerical data. We report on recent progress related to analytical finite volume studies in case of the pion mass.

A long time ago Lüscher established an asymptotic formula [1] which relates the size dependence of the pion mass M_π with the $\pi\pi$-forward scattering amplitude $\mathcal{F}_\pi(\nu)$,

$$M_{\pi L} - M_\pi = -\frac{3}{16\pi^2 M_\pi L} \int_{-\infty}^{\infty} dy\, \mathcal{F}_\pi(iy)\, e^{-\sqrt{M_\pi^2+y^2}L} + O(e^{-\sqrt{3/2}M_\pi L}) \ , \quad (1)$$

with $M_{\pi L}$ the pion mass in finite volume and where the constraint $M_\pi L \gg 1$ is assumed. The contributions in the low energy region are enhanced, where the $\pi\pi$-scattering amplitude is represented very accurately through its chiral representation [2, 3]. At leading order in the chiral counting the forward scattering amplitude assumes to be a constant, and the integral simplifies to a modified Bessel function of the second kind. When Colangelo and Dürr evaluated subleading finite volume effects with Lüscher's formula, they observed numerically enhanced corrections with respect to the leading order [4]. These investigations show a clear necessity to go beyond leading order calculations in order to have the finite volume effects of the pion mass under control. Furthermore, as the derivation of eq. (1) takes into account only exponential contributions of the order $\exp(-M_\pi L)$ and systematically drops those of the order $\exp(-\sqrt{3/2}M_\pi L)$, the question arises, whether the missed terms might turn out to be numerically relevant. The same question also concerns the asymptotic formulae of decay constants [5]. As we will show, a full two-loop calculation of the pion mass in finite volume ChPT clarifies these open points.

The two-loop calculation appears to be interesting also for its own right. To date, a number of finite volume calculations have been performed at one-loop order [6], but as far as we know two-loop calculations have only been performed for the quark condensate [7] and for low-energy observables in a closely related field [8]. As finite volume effects occur first at the one-loop level, only a two-loop calculation – or alternatively if existing, an asymptotic formula à la Lüscher – leads to a better understanding of the convergence behaviour of the perturbative expansion.

2 ChPT in finite volume

Chiral Perturbation Theory (ChPT) is the effective theory for QCD at low energies. It is nowadays a mature field which has been applied successfully for a variety of phenomena, in particular in the meson sector. For an introduction and a current status of the field, we refer to Ref. [9, 10].

The effective framework is still appropriate, when the system is enclosed by a box of size $V = L^3$. We refer to the literature for the foundations [11, 12, 13] and a recent review [14]. Here, we only remind of the fundamental results which guided the present calculation: the volume has to be large enough, such that ChPT can give reliable results, $2F_\pi L \gg 1$. The perturbative calculation is bound to the value of the parameter $M_\pi L$. Whether it exhibits to be large ($M_\pi L \gg 1$, "p-regime") or small ($M_\pi L \lesssim 1$, "ϵ-regime") implies a different power counting. However, in both cases the effective Lagrangian is the same as in the infinite volume. Here, we only cover the "p-regime", where the system is distorted mildly and the only change brought about by the finite volume is a modification of the pion propagator due to the periodic boundary conditions of the pion fields

$$G(x^0, \mathbf{x}) = \sum_{\mathbf{n} \in \mathbb{Z}^3} G_0(x^0, \mathbf{x} + \mathbf{n}L), \qquad (2)$$

with $G_0(x)$ the propagator in infinite volume.

3 Pion in finite volume

The pion mass in finite volume is defined by the pole equation

$$G(\hat{p}_L)^{-1} = 0, \qquad \text{for} \qquad \hat{p}_L = (iM_{\pi L}, 0), \qquad (3)$$

where $G(p^0, \mathbf{p})^{-1}$ is the Fourier transform of the connected correlation function

$$\langle \pi^1(x)\pi^1(0)\rangle_L = L^{-3} \sum_{\mathbf{p}} \int \frac{dp^0}{2\pi} e^{ipx} G(p^0, \mathbf{p}),$$

$$G(p^0, \mathbf{p})^{-1} = M^2 + p^2 - \Sigma_L(p^2), \qquad (4)$$

with $\Sigma_L(p^2)$ the self-energy in finite volume and M^2 the pion mass in the chiral limit in infinite volume. A determination of the pion mass amounts thus to an evaluation of the self-energy in a loop expansion. At one-loop order the finite volume corrections have been evaluated in [11]. Here, we discuss the main steps which guided the two-loop calculation. A detailed derivation of the results will be given elsewhere [15]. It is convenient to write the pion mass in finite volume to two loops in the following manner,

$$\begin{aligned} M_{\pi L}^2 &= M_\pi^2 - \Sigma^{(1)} - \Sigma^{(2)}, \\ M_\pi^2 &= M^2 - \Sigma^{(0)}, \end{aligned} \qquad (5)$$

where $\Sigma^{(r)}$, $r = 0, 1, 2$, denote the contribution of the self-energy with r propagators in finite volume with non-vanishing vector $\mathbf{n}$ (cf. eq. (2)). These terms shall be discussed in some detail.

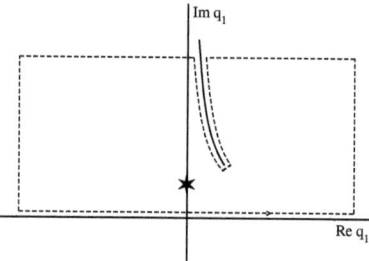

Fig. 1. Integration contour in the complex q_1 plane (dashed line) with the pole from the pion propagator (diamond) and the branch cut from the $\pi\pi$-scattering amplitude.

3.1 Self-energy to 0'th order: $\Sigma^{(0)}$

The contributions to $\Sigma^{(0)}$ are volume independent by definition and merely renormalize the pion mass, cf. eq. (5). A detailed discussion of this calculation can be found in [2], with which we agree.

3.2 Self-energy to 1'st order: $\Sigma^{(1)}$

As Lüscher showed [1], the leading finite volume effects are captured in $\Sigma^{(1)}$ which can be summed up in closed form

$$\Sigma^{(1)} = \frac{1}{2} \int \frac{d^4q}{(2\pi)^4} \sum_{n=1}^{\infty} m(n)\, G_0(q) e^{iq_1\sqrt{n}L}\, \Gamma_{\pi\pi}(\hat{p}, q, -\hat{p}, -q)\,, \qquad (6)$$

with $\hat{p} = (iM_\pi, 0)$, $\Gamma_{\pi\pi}(\hat{p}, q, -\hat{p}, -q)$ the 4-point function of $\pi\pi$-scattering in the forward scattering kinematics and $m(n)$ the number of integer vectors z with $z^2 = n$. Eq. (6) may still be simplified considerably. We perform a contour integration in the complex plane of the first component q_1. Singularities are met by the pole of the propagator as well as the branch cuts from the propagator and the 4-point function. At the two-loop level only the cuts of the 4-point function have to be considered, lying on a hyperbola in the complex q_1 plane, as illustrated in fig. 1. Performing the contour integration along the dashed line, we obtain two terms, one from the residuum of the pole I_p and the other from the integral along the new integration path. The latter contribution vanishes as we push the integration lines to infinity, except for the one along the cut, to be denoted by I_c in the following,

$$\Sigma^{(1)} = I_p + I_c\,. \qquad (7)$$

Along the lines discussed in [1], the former leads to the resummed asymptotic formula

$$I_p = \frac{M_\pi^2}{16\pi^2 \lambda_\pi} \sum_{n=1}^{\infty} \frac{m(n)}{\sqrt{n}} \int_{-\infty}^{\infty} dy \, \mathcal{F}_\pi(iy) \, e^{-\sqrt{n(1+y^2)}\lambda_\pi} \,, \tag{8}$$

with $\mathcal{F}_\pi(iy)$ the $\pi\pi$-forward-scattering amplitude. Restricting the sum to the first addend, we recover Lüscher's formula, eq. (1). Its extension to eq. (8) has already been suggested in [14, 16]. Numerically, the resummation turns out to be relevant for moderate λ_π.

Concerning the contributions of the cut I_c, we make use of a dispersive treatment. As these terms only start at the two-loop level, they are both suppressed in the chiral as well as in the large L expansion and are therefore expected to be small. This is exactly what we observe numerically.

3.3 Self-energy to 2'nd order: $\Sigma^{(2)}$

Contributions from two pion propagators in finite volume are finally captured in $\Sigma^{(2)}$. Notice that the corresponding Feynman diagrams are uv-finite and need not to be renormalized. Therefore, it only remains to find a feasible numerical representation for these terms.

4 Numerics

The numerical analysis is performed in line with the setup of Ref.[16]. In fig. 2, we evaluate the relative finite volume shift

$$R_{M_\pi} \equiv \frac{M_{\pi L} - M_\pi}{M_\pi} \,, \tag{9}$$

for $L = 2, 3, 4$ fm as a function of M_π. We show the one-loop (LO) as well as the two-loop result (NLO). These shall be compared with the resummed asymptotic formula eq. (8) with LO/NLO/ NNLO input for the $\pi\pi$-scattering amplitude. Note that the one-loop result and the resummed asymptotic formula to LO agree with each other. The best estimate for R_{M_π} is finally obtained by adding to the asymptotic pure three-loop contribution the two-loop result (NNLO asympt. + NLO non-asympt.). At NLO, the finite size effects encounter low energy constants,

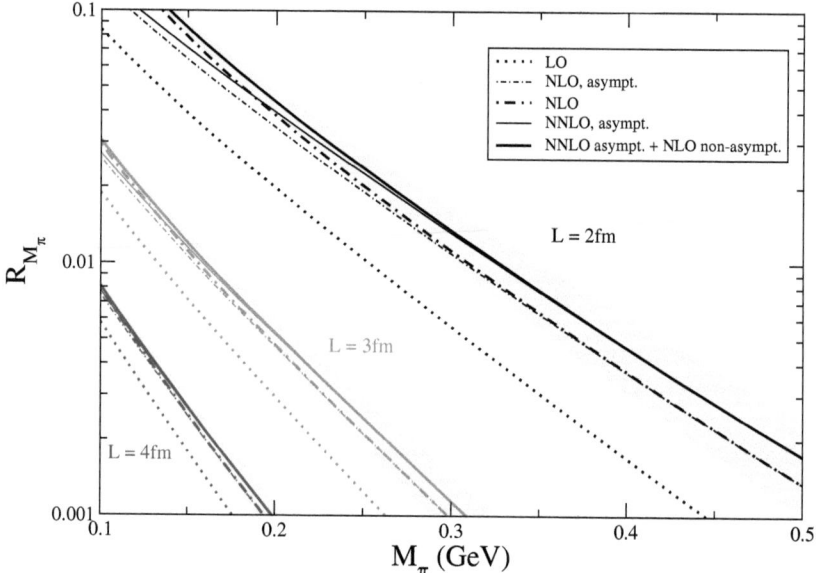

Fig. 2. $R_{M_\pi} = M_{\pi L}/M_\pi - 1$ vs. M_π for $L = 2, 3, 4$fm. For explanations of the legend, see text.

leading to a non-negligible error band which is only shown for the best estimate. We take up a point already alluded in the introduction, namely the large contributions when going from LO to NLO in the asymptotic formula (dotted to thin-dash-dotted). Compared to this gap, the additional contributions from the full two-loop result (thick-dash-dotted) are very small. Consider eg. a pion mass of $M_\pi = 250$ MeV in a 2 fm box. We find $R_{M_\pi} = 0.0236(41)$, of which 0.0010 stem from the two-loop corrections which are not included in the asymptotic formula. The two-loop and the NLO result from the asymptotic formula only drift away, when we go beyond the region where the p-regime can be safely applied.

5 Conclusions

We have evaluated the finite volume effects of the pion mass to two loops within ChPT in the p-regime. The results are compared with a recently proposed extension of the asymptotic formula of Lüscher. We find that contributions which were neglected in the latter, are numerically very small at the two-loop level and conclude that the resummed asymptotic formula is a convenient method to evaluate the finite volume effects beyond the leading order.

Acknowledgments

The work presented here is being done in collaboration with Gilberto Colangelo whom I warmly thank, also for a careful reading of the manuscript. This work is supported by the Swiss National Science Foundation and in part by RTN, BBW-Contract No. 01.0357 and EC-Contract HPRN–CT2002–00311 (EURIDICE).

References

[1] M. Luscher, Volume Dependence Of The Energy Spectrum In Massive Quantum Field Theories. 1. Stable Particle States, Commun. Math. Phys. 104, 177 (1986).

[2] J. Bijnens, G. Colangelo, G. Ecker, J. Gasser and M. E. Sainio, Pion pion scattering at low energy, Nucl. Phys. B 508, 263 (1997) [Erratum-ibid. B 517, 639 (1998)] [hep-ph/9707291].

[3] G. Colangelo, J. Gasser and H. Leutwyler, pi pi scattering, Nucl. Phys. B 603, 125 (2001) [hep-ph/0103088].

[4] G. Colangelo, S. Durr and R. Sommer, Finite size effects on M(pi) in QCD from chiral perturbation theory, Nucl. Phys. Proc. Suppl. 119, 254 (2003) [hep-lat/0209110]. G. Colangelo and S. Durr, The pion mass in finite volume, Eur. Phys. J. C 33, 543 (2004) [hep-lat/0311023].

[5] G. Colangelo and C. Haefeli, An asymptotic formula for the pion decay constant in a large volume, Phys. Lett. B 590, 258 (2004) [hep-lat/0403025].

[6] A. Ali Khan et al. [QCDSF-UKQCD Collaboration], The nucleon mass in N(f) = 2 lattice QCD: Finite size effects from chiral perturbation theory, Nucl. Phys. B 689, 175 (2004) [hep-lat/0312030]. S. R. Beane, Nucleon masses and magnetic moments in a finite volume, Phys. Rev. D 70, 034507 (2004) [hep-lat/0403015]. S. R. Beane and M. J. Savage, Baryon axial charge in a finite volume, Phys. Rev. D 70, 074029 (2004) [hep-ph/0404131]. D. Becirevic and G. Villadoro, Impact of the finite volume effects on the chiral behavior of f(K) and B(K), Phys. Rev. D 69, 054010 (2004) [hep-lat/0311028]. D. Arndt and C. J. D. Lin, Heavy meson chiral perturbation theory in finite volume, Phys. Rev. D 70, 014503 (2004) [hep-lat/0403012].

[7] J. Bijnens, N. Danielsson, K. Ghorbani and T. Lahde, Two Loop Partially Quenched and Finite Volume Chiral Perturbation Theory Results, [hep-lat/0509042].

[8] A. Schenk, Pion propagation at finite temperature, Phys. Rev. D 47, 5138 (1993). D. Toublan, Pion dynamics at finite temperature, Phys. Rev. D 56, 5629 (1997) [hep-ph/9706273].

[9] G. Colangelo and G. Isidori, An introduction to CHPT, [hep-ph/0101264]. J. Gasser, Light-quark dynamics, [hep-ph/0312367]. S. Scherer and M. R. Schindler, A chiral perturbation theory primer, [hep-ph/0505265].

[10] U. G. Meissner, H. W. Hammer and A. Wirzba, Chiral Dynamics: Theory and Experiment (CD2003), [hep-ph/0311212].

[11] J. Gasser and H. Leutwyler, Light Quarks At Low Temperatures, Phys. Lett. B 184, 83 (1987).

[12] J. Gasser and H. Leutwyler, Thermodynamics Of Chiral Symmetry, Phys. Lett. B 188, 477 (1987).

[13] J. Gasser and H. Leutwyler, Spontaneously Broken Symmetries: Eeffective Lagrangians At Finite Volume, Nucl. Phys. B 307, 763 (1988).

[14] G. Colangelo, Finite volume effects in chiral perturbation theory, Nucl. Phys. Proc. Suppl. 140, 120 (2005) [hep-lat/0409111].

[15] G. Colangelo and C. Haefeli, to be published

[16] G. Colangelo, S. Durr and C. Haefeli, Finite volume effects for meson masses and decay constants, Nucl. Phys. B 721, 136 (2005) [hep-lat/0503014].

Acknowledgments

The reason why I enjoyed the last three years so much is easily accounted for: it is due to my supervisor Gilberto Colangelo. I could not have imagined having a better advisor and mentor for my PhD. His wise, friendly and humorous nature has impressed me deeply and influenced this thesis fruitfully. I would also like to express my gratitude to Stephan Dürr whose insights were always very valuable and to Rainer Sommer for kindly agreeing to be coreferee of this thesis. My thanks go also to all the (former) members of the institute, especially to Ruth Bestgen, Kay Bieri, Thorsten Ewerth, Andreas Fuhrer, Ottilia Hänni, Florian Kämpfer, Bastian Kubis, Markus Moser, Emanuel Nikolidakis, Martin Schmid, Julia Schweizer and Peter Zemp who created an enjoyable atmosphere. To my parents and my friends I am deeply indebted for their love and support during my studies. Brigitte, thanks for being there!

This work is supported by the Swiss National Science Foundation and in part by RTN, BBW-Contract No. 01.0357 and EC-Contract HPRN–CT2002–00311 (EURIDICE).

Die VDM Verlagsservicegesellschaft sucht für wissenschaftliche Verlage abgeschlossene und herausragende

Dissertationen, Habilitationen, Diplomarbeiten, Master Theses, Magisterarbeiten usw.

für die kostenlose Publikation als Fachbuch.

Sie verfügen über eine Arbeit, die hohen inhaltlichen und formalen Ansprüchen genügt, und haben Interesse an einer honorarvergüteten Publikation?

Dann senden Sie bitte erste Informationen über sich und Ihre Arbeit per Email an *info@vdm-vsg.de*.

Sie erhalten kurzfristig unser Feedback!

VDM Verlagsservicegesellschaft mbH
Dudweiler Landstr. 99   Telefon +49 681 3720 174
D - 66123 Saarbrücken   Fax   +49 681 3720 1749
www.vdm-vsg.de

Die VDM Verlagsservicegesellschaft mbH vertritt

Printed by Books on Demand GmbH, Norderstedt / Germany